PHILOSOPHERS STONE

MAKE THE WORLD A BETTER PLACE

HENRY KROLL

DISCLAIMER
The information presented in this book is meant for scientific and entertainment purposes only. The author makes no express claims that the recipes or information will cure disease or illness of any kind. Differences in physiology, DNA and body type make this impossible to predict. At the first sign of illness or malady of any kind a person should seek the advice of a doctor or medical professional.

ISBN: 979-8-89419-542-1 (sc)
ISBN: 979-8-89419-543-8 (hc)
ISBN: 979-8-89419-544-5 (e)

Because of the dynamic nature of the Internet, any web addresses or links contained in this book may have changed since publication and may no longer be valid. The views expressed in this work are solely those of the author and do not necessarily reflect the views of the publisher, and the publisher hereby disclaims any responsibility for them.

One Galleria Blvd., Suite 1900, Metairie, LA 70001
(504) 702-6708

CONTENTS

Preface v

Chapter 1   The White-Powder-Of-Gold.................................... 1
Chapter 2   Transmutation of Elements 7
Chapter 3   Reduce Stress Inflamation, Be Happy and Live Longer 12
Chapter 4   Vaccines 21
Chapter 5   Big Threats to All Life on Earth 28
Chapter 6   The Coronavirus Simulation................................. 33
Chapter 7   Biggest Threat to Freedom 41
Chapter 8   How the Gods Created Man.................................. 48
Chapter 9   Enki and the Moon 65
Chapter 10   Et's and Non-Local Visitations............................. 87
Chapter 11   Today's Mind Controll.. 97
Chapter 12   David Hudson 112
Chapter 13   The Ark is a Brain-Wave Transmitter and receiver 125
Chapter 14   Gold of the Gods 143
Chapter 15   Morphogenetic Fields and Intelligent Plasma................. 152
Chapter 16   Transmutation How to Make Gold........................162
Chapter 17   Antigravity...172
Chapter 18   How to Live 900 Years the Blessing Way 175
Chapter 19   Cancer Cures Where to Look for the Gold of the Gods... 182
Chapter 20   Ethical and Philosophical Considerations 188
Chapter 21   How to Make the Philosopher's Stone....................211

About the Author.. 257
Addendum 261

PHILOSOPHER STONE

The philosopher stone or manna—also known as the sacred, Hebrew MFKZT powder or white-powder-of-gold was carried in the Ark of the Covenant. If you were to peek inside you would see a glowing white powder with little lightning bolts trickling over the surface. This antigravity, room-temperature, superconductor is used for cell division by all living things including plants and was used by God's angels to create man.

David Hudson spoke in a lecture, "The manna or the elixir of life when mixed with water forms a gelatinous mixture. When ingested it will have the following effects, every sell in your body will be taken back to the state it is supposed to be, when you were a teenager or a child. It perfects the DNA and closes the light within the body until you literally reach a point where the light body exceeds the physical body.

"The gifts that go with it are perfect telepathy, you can know good and evil when it in the room with you, you can project your thoughts into someone else's mind, you can levitate, you can walk on water, because it is flowing so much light in you, you literally don't attract to gravity."

Teams of scientists in Japan, Germany, China, and the US have been experimenting with the technology since the1980's. It was recently discovered in the air we breathe and the water we drink. The air filters which you plug into your home to collect dirt, dust and pollen also, collect ORME atoms. It looks like a dirty-white gel. When you touch it with your finger it disappears because the magnetic field of your finger puts it into a higher energy state. This may be the substance that answers the Parana energy mystery. Athletes can lift more and run faster after taking a deep breath. It can't be the oxygen intake because oxygen takes several minutes to get into the blood stream. The human olfactory nerves in the nasal passage have the ability to absorb ORME atoms.

This material super-conducts electricity during cell division and corrects the DNA in all living things. It can increase your life-span many years and increase your IQ a hundred points or more. It is the only known room-temperature superconductor. Due to it's physical shape like a solar system or galaxy it repels the force of a magnet! It improves electricity production in fuel cells and solar cells.

When used to bring for the shekinah glory (electric discharge of holy fire) over the top of the Ark a signal can be broadcast across the galaxy in one Planck second. You don't have to wait millions of light years for an answer when communicating with other intelligent races in our galaxy. The signal is sent via magnetic frequency not electromagnetic frequency. It is not a radio frequency. It is a higher dimensional frequency created by the Cooper-paired, electrons passing each other at twice light speed. If you tripled your intelligence you could instantaneously produce holographic thought-forms images above and identical Ark on the other side of the Galaxy on whatever planet this device happens to be operational.

Knowledge or the ORME (Orbitially Rearranges Monatomic Elements) particle which Moses metal smith, Bazaleal created by burning the golden idol at 5,600 degrees is a little difficult for people to understand. The concept of a monatomic particle that cannot bond into a solid substance and has to float around is because it's electrons are orbiting in pairs. It is a particle with a magnetic frequency of brain waves enabling the observer to manipulate the outcome of an experiment or event. It is the frequency that broadcast morphogenetic fields allowing flocks of birds to fly without colliding. Animals and certain humans can know things before they happen. ORMUS makes mental telepathy possible. This advanced, alien-mind-control science is a little difficult for the average person to comprehend.

It exists, as a glowing white powder in its pure state with thousands of amps of electricity flowing through it but no voltage unless excited by brain wave frequency or some other outside frequency. This is the primal atom used to regulate the outcome of a singularity during the creation of the Universe.

It weight five-eighths the weight of the actual metal itself at room temperature but when heated weighs less than nothing and disappears from third dimension. When it cools it reappears again from a higher dimension.

The ancient Egyptian translation of manna is "What is it?" It is the food of the Gods and was used to take mankind into higher dimensions where time and distance are inconsequential. This has been done since the beginning of time.

You eat it in your food every day yet most people are completely unaware that it exists. People pray over their food to energize it so that it will nourish their body more completely and this material actually responds to those prayers by altering the frequency thereby making the food more compatible and digestible. It is the light of life and absolutely necessary for cell division in all living things. Plants take it out of the soil by dissolving it with alkaline extruded from their roots. You break the plants down using hydrochloric acid in your stomach. The reason why some plants are poison is because they contain too much alkaline. Most plant poison are alkaline poisons.

The use of this material makes the cloning of all life forms possible including human eugenics. Moses fed the Hebrew children the priest bread laced with the whit-powder-of-gold, the sacred MFKZ powder. This act was against the law. The priests and Pharaohs kept this technology secret for corruption and power. It corrected the Hebrew children's DNA making them smarter and stronger than their half-brothers Arabs who all came from the seeds of Abraham. When used as such it gives mankind the ability of a God to speed up evolution.

Such power is, I feel in accord with God's will especially when used for a worthwhile, cause such as curing disease and getting us off the planet to assist and create life in other biospheres. The use of this material will reduce the stress on our own biosphere. We must take the next evolutionary step or perish.

There are certain drawbacks to taking the stuff however because you suddenly go from using eight percent of your brain to maybe eighty percent. The shock of such a transformation is enough to kill most people.

I debated with myself several years before including the recipes in this book because of the potential harm they might cause. Finally, I thought to myself, "This information had been kept secret long enough. We can't shield mankind from this knowledge forever." Enclosed are several recipes which you can use in your kitchen to extract it from sea water or transmute it from Dead Sea minerals, sea salt or pure gold.

According to the American Indians we are now living in the fifth world. This means there were four other advanced civilizations before us which were exterminated either due to natural events or the downright stupidity of atomic war. It is tantamount that we develop space travel to leave a historical record on other planets. If we continue to plod along at our present pace of scientific advancement there may be no one left to tell the story.

For more information on this subject red my book on astronomy title: COSMOLOGICAL ICE AGES. Also check our Space Ships of the Gods and Home of the Angels. www.HankKroll.com

CHAPTER ONE

THE WHITE-POWDER-OF-GOLD

This is the second printing of Philosophers Stone. I wrote the original version about thirty years ago. Now at age 77 I have accumulated more incite about ageing and how to build your immune system to protect yourself from disease. The problems we have today with so many immune diseases and inflammation is that we eat supermarket food grown on the same soil for 200 years. It is practically devoid of trace minerals.

PHILOSOPHERS STONE is about bypassing the plant to feed your cells with the proper nutrients for complete cell division. Instead of eating the plants or meat you take the trace minerals directly out of soils with alkaline solutions. It's the same process that plants use. Plants exude alkaline solutions from their roots to dissolve the minerals and nutrients into solution. This mineral rich solution is carried up into the leaves where the plane uses light to make ATP. This process has been around since the invention of plants.

You simply boil mineral rich soils or sea salt and other source items in alkaline water to get them into solution thus speeding up the long process of waiting for plants to grow.

After you have located a good clean source of minerals you put the minerals in your bread or other foods. The trace minerals can also be put in your drinking water. The point is, you need to feed the mitochondria in each of your human cells to slow the ageing process and correct DNA defects.

The soils around the Phoenix Arizona and the Middle East were covered with oceans at one time. When the sea water evaporated it left all the minerals and orbitally rearranged atoms of gold and platinum group and other minerals in the soil. You can't grow anything on these soils because they are too alkaline. Plants can't grow there. Modern farmers are lowering the PH of desert soils with acid.

David Hudson who farms west of Phoenix, Arizona buys 90 tons of sulfuric acid from Freeport Moran and sprays it on five acres. After five years he plants cotton on it and it grows fine. After that he can plant tomatoes and they grow wonderfully. It's all a matter of understanding PH.

Some, not all deserts were covered with sea water at one time. When the water evaporated it left all the elements that were in the water in the soil. Another way to find a natural deposit of ORMUS concentrated trace minerals look for a dry-wash where the infrequent rains in the deserts of the southwest United States have cut a ditch down through the soil. If you are lucky you see a white band of material about three or four inches wide about two or three-feet down below the desert floor. The whitish band of material is the trace elements that have leached down through the soil over the last several thousand years.

There would be a high concentration of magnesium along with all the trace elements that were in the ancient oceans in that band of white material. You can scoop it out with a spoon and eat it. Magnesium is the same stuff that is in milk of magnesia used to cure stomach ailments. You can purify it yourself using the recipes in this book.

When you're bypassing the plant you don't have to wait several months for your garden to grow. The ancient peoples take the desert soil and boil it in the alkaline desert spring water for five days or more. After five days the liquid becomes a deadly poison because they had to keep adding the alkaline spring water.

The clear liquid is poured out and the soil thrown away. It now has all the beneficial nutrients and trace elements that were in the soil only you can't see them. If you drank this liquid it would burn your mouth and you might die from alkaline poisoning. All your hair would fall out. The alchemists reduced the PH by adding an acid like vinegar or whine and suddenly the atoms become visible in solution as milk. If they reduced the PH enough they could drink it.

The ancient people filtered it with fine cloth and put in the Ark of the Covenant which is a box designed to exclude magnetic fields with layers of insulating wood and metal. One had to be careful not to expose the monatomic ORMES in direct sunlight because it can catch fire and burn like magnesium metal. When it burns it gives off beta radiation which is dangerous to anyone standing nearby.

The purified trace elements and ORMES taken out of the soil is the manna carried in the ARK of the covenant. I will get into the spooky properties of this inter-dimensional material later in the book. Einstein called it spooky communication because like splitting two light beams and altering one the other receives the same information instantly much faster than the speed of light. Due to its paired electrons dipping out of hyperspace it is inter-dimensional transcending all time and space.

The recipes in this book were developed by Pfizer Pharmaceutical to take the minerals out of Dead Sea salt, sea water, granite powder and limestone that was grown in shallow seas by plants. The technology has been around thousands of years but kept secret by the rich and powerful to extend life and increase intelligence.

At the present time most scientists believe that transmutation of elements is impossible because they believe it takes millions of electron volts to pull atoms apart to turn them into gold or other elements.

Yes, it's true. Atomic forces within atoms are very strong making it impracticable due to the high cost of energy to change one element into another. However, when dealing with ORMUS or orbitally rearranged monatomic elements also known as the sacred Hebrew MFKZT powder are combined atoms with all the protons in one giant nucleus. Therefore ORMES or M-state has weak atomic forces. With double the number of protons in a single nucleus they have weak atomic forces because protons are like little magnets of the same polarity. They push apart making it possible to pull protons out with little magnetic energy pulses.

If you feed the mitochondria the proper nutrients then they can do a perfect job of perfecting the DNA. They use magnetic frequency to rip out protons to create the necessary e4lements for complete cell division without losing any DNA segments and telomeres at the end of the DNA chains. This is where science doesn't get it because they believe transmutation of elements is impossible.

The mitochondria pull protons out of ORMUS with magnetic pulses of high frequency to make any element necessary for complete cell division. However if they don't have enough of the large combined atomic particles or the right trace minerals they can't do their job and you lose body cells at a faster rate. This is why some people afe faster than others.

MITOCHONDRIA

The mitochondria you inherited from your mother are in the egg at conception. They have their own DNA. Extracting mitochondria from human cells left at a crime scene can be used in court to identify the perpetrator and race of the individual. Everyone's DNA is unique but similar to their parents and siblings.

Scientists from all over the world have been studying mitochondrial DNA for four decades. So far they have identified more than 48 types of mitochondria DNA. More recently the human genome consisting of three-billion RNA and DNA segments has been cataloged so that computers can identify who were your ancestors and what part of the world they originated.

The replication of body cells is a numbers game. In order to slow the ageing process or remain the same age you have to replicate 6 to 8-million body cells every second. Of course this takes a lot of energy. You have to feed them the proper elements and trace mineral so that they can do their job.

ORMES or ORMUS exists everywhere. It's in the air you breathe, and in the food you eat. Fish and seaweed and almost anything grown in the sea is contains more ORMES than supermarket food. However you have to consider how much pollution and heavy metals are in the area. Be careful not to eat fish caught near large manufacturing facilities that might be dumping toxic waste.

If you could see one ORME it would look like a small galaxy. All the electrons whizz around in opposite directions known as cooper pairing. Because they are ring shaped their physical shape repels magnetic fields. They fit the technical definition of a room-temperature, super conductors in that they reproduce magnetic fields with no loss of power.

This book contains some of the history and use of the white-powder-of-gold as it's known today ORMUS or ORMES, (orbitally rearranged monatomic elements). Nowadays you can order it off the internet. Youtube has hundreds of videos on the subject.

AGEING

The main cause of ageing is loss of body cells. The rate of cell replication process is determined by the number of turns of DNA. When you are conceived in the womb you have 50 base pairs per turn. After you are born the cell division slows down and you have roughly 25 base pairs per turn of DNA. Age 20 is the breakeven point where you have 20 base pairs per turn of DNA. If you can maintain the 20 base pairs per turn of DNA by feeding the mitochondria with the right nutrients theoretically you can live forever.

At age fifty the number of base pairs per turn of DNA is only 5 and you start losing body cells exponentially. I didn't mention the telomeres which are little monticule chains at the end of the gene code. They are like little clocks that get shorter and shorter each time the cell replicates. When the telomeres are gone the cell dies. Ageing is caused by loss of body cells.

The job of the mitochondria is to furnish enough energy to replicate their own cell DNA and enough energy to replicate your human cell DNA as well at the same time. If you don't feed them properly they can't do their job correctly and you lose DNA segments and body cells.

There is however good news. If you feed the mitochondria the right nutrients they can actually correct DNA defects and rebuild the telomeres and actually wind the DNA up tighter thus increasing the cell replication rate. Apparently the ancient man of renown mentioned in the Bible knew these secrets leaned from their creators because many of them lived hundreds of years, had many wives and begat fifty or more children to populate the earth.

The problem we have today is that most of the food we eat is food grown on the same soil for 200 years. Most all the trace minerals have been used up a long time ago. To increase production they use chemical fertilizer and pesticides that can accumulate in your body over time causing cancer and a host of other maladies. What if you could supplement your diet with abundant trace minerals and ORMES like the pharaohs and priests then you can live hundreds of years.

CHAPTER TWO

———————

TRANSMUTATION OF ELEMENTS

Transmutation of elements is happening all around us in all living things but our schools teach us that it doesn't exist. School curriculums are selected by government overseer panels hired by the Rockefeller Foundation who want a slave race of people to serve them. They want to keep us stupid so they use every trick in the book. The globalist agenda is to create a population with just enough education to work at a mundane job. They especially don't want us to live long because it might bankrupt Social Security. They want young productive workers to serve the elite and die soon after our most productive years are used up they want you dead so the government can keep the Social Security money you paid in. They especially don't want a population that is smarter than they.

The Rockefeller Foundation working with the faculty of Harvard University also laid the foundation for our present day medical practice where alleged "doctors" experiment on you and call it "practicing medicine." That's exactly what doctors do. They practice on you. In other words you are a 6-foot Guiney pig that was dumbed down by the school system to where you actually believe the medical profession can cure disease.

The Rockefeller ruling elite 'agenda' harkens back to the Middle-Ages where the Dukes, Earls, Princes, Barons and Kings owned the land and the people on it. The ruling elite stood a foot taller than the average person and lived much longer to rule like kings. They lived so much longer than the average person because they had the technology and a better diet than the uneducated peasants.

They needed several wives and concubines with fifty or a hundred children to create an army of young men to defend their way of life. Most lived in stone castles with a draw bridge and a moat to protect themselves from invaders. They fortified their castles with cannons and vats of boiling oil. They had an army of large men dressed in armor ready to fend off all invaders.

The only way you could conquer them was to poison their wells and starve them to death. This could take several years because they had large grain storage bins built into their castles. Catapults were sometimes used to throw flaming bails of straw soaked in oil over the castle walls to smoke them out.

The poor servants and farmers living outside the walls of the castle had no way to defend themselves. In some cases the elderly were killed and the young women and boys were taken to serve the new masters.

The common people lived in huts made of mud and sticks. To keep from freezing to death in winter they brought the cow inside—that is, if they were fortunate enough to have a cow. Most all livestock and land belonged to the Barons, Dukes, Earls and Kings. This is the kind of lifestyle Bernie Sanders has planned for our future here in America.

HOW THEY KEEP US STUPID

A fleet of large tanker airplanes currently known as Evergreen Air and formerly known as Air America was created to spray agent-orange and drop napalm and other horrible things like nerve agents on Vietnam. These were war crimes. They had a DC3 equipped with a machine gun that could spray thousands of rounds a second covering every square foot of ground with a bullet. They called it, "Puff, the Magic Dragon" that could fly over an area with a machine gun that could spray bullets so fast that it could put a bullet in every square yard below the plane. Why weren't the people responsible for these crimes punished? There all dead now so what's the point?

EVERGREEN AIR

After the war the huge tanker planes and Douglas DC3s and 6es were brought back to America and based in Oklahoma to spray parquet on the marijuana fields over the border into Mexico. Mexican drug lords were making too much money selling the deadly marijuana to Hippies who had protested the Vietnam War.

This same fleet of airplanes now sprays aluminum powder, boron and other chemical waste over the major cities and the continental United States to further dumb down the population. Their excuse for spraying chemical trails is to stop the mythical "global warming."

Whenever there is a perceived threat to our national security such as a small nation getting their hands on a nuclear bomb we have to act. The United States military asks the government for more money to go in there and occupy the place. We have 96 military bases placed all over the world plus embassy's and staff to protect them. Many of these occupations last decades and some forever. The US is like a big mean dog. Don't wake it up!

Instead of going in and bombing the hell out of a place and killing and destroying the infrastructure like we did in Iraq so that the women and children suffer horribly maybe it's better in some cases like Cuba to let the despots and dictators rule. Eventually they get fat and lazy and die of old age. This process of regime change has been going on for thousands in many countries around the world where the leaders become complacent and a more aggressive political faction takes over. One could call this process evolution of the species.

BIG CORPORATIONS ARE INVOLVED IN IT TO MAKE MONEY

Believe it or not, Monsanto and Archer Daniel Midland Company or ADM have developed seeds that can grow on aluminum contaminated soil. The same corporate giants also developed seeds that will only grow one time thus forcing farmers to buy all their seeds from them; the ultimate corporate evil. Farmers in India have been committing suicide and their families are starving because they can't grow their own seed and they can't afford to buy seed from ADM or Monsanto.

Our rulers want a docile herd dumbed down with mercury in the vaccines, fluoride in the toothpaste and in the re-cycled drinking water. They want us to work hard every day to provide goods and services so they can live like kings. Honestly I am amazed the ruling elite haven't made more attempts on my life for releasing this information. Maybe the subject matter is too complicated for them to understand.

OUR GOVERNMENT TURNS PEOPLE INTO
14TH AMENDMENT SLAVES AT BIRTH

When you are born your mother is tricked into signing the BIRTH CERTIFICATE. The BIRTH CERTIFICAT is written on BOND paper. It is a real bond. If you take a magnifying glass and read the very fine print on the bottom you will see the name of the company that made the bond paper. You will also find a red bond number on that document. The bonds are sent to the Bureau of Vital Statistics located in the pyramids structure building across the street from the New York Stock Exchange in New York. You can look up your bond number to see how valuable it is.

Birth bonds are what back the money in this country. Gold and silver no longer is tied to the dollar. Roosevelt took us off the gold standard in 1933 and the government confiscated all the gold including gold coins. My poor mother had a ten dollar gold piece she kept all through World War II. She worried that the government would find out about it.

When a baby is born the hospital used to get $5,000 from the government for each child born in their facility. It's probably more like $8000 now. Human lives are worth over ten-million dollars. Whenever people are killed in a war or mass shooting the Generals and news media refer to it as collateral damage. That's exactly what we are. We are collateral. All future children born in this country are used as collateral to back the currency with their future production or labor.

BIRTH BONDS re bundled together and sold on the New York Stock exchange. The Franklin Templeton Fund and others buy these bonds as investments because they increase in value over time.

When you are charged with a crime the COURTS form a non-profit trust to take money out of your BIRTH BOAND. That's why for certain crimes they can fine you and charge you more or less depending on the severity of the crime. They charge your corporate entity that is written in all CAPITOL LETTERS on your BITRH BOND and on the papers you are charged with. When you "sign" your name next to your name in all CAPITOL LETTERS you authorize them to take money out of your 'CORPORATE BOND'. The judge has a list of crimes and the amount to charge for each crime in front of him when he makes his final decision.

EXAMPLE OF A 'PERSON' WHO DIDN'T HAVE A BIRTH CERTIFICATE BOND

They couldn't charge him because he was worthless to the COURT. There was no money for the COURT to take from his BIRTH BOND because he didn't have one. The definition of a 'PERSON', is a non-living fiction corporation created by the government with an all CAPITOL LETTER NAME!

A Native Alaska friend of mine was accused of attempted murder by his fellow board members because he objected to them stealing Native Corporation money. They accused him of trying to run one of them over with his pickup truck when he was exiting the parking lot. The whole case was a set-up to put him in jail.

When the Judge demanded that he show some kind of identification he brought out a piece of birch tree bark that his mother had written documenting when and where he was born. Because he didn't have a bond on his head the court couldn't charge him. There was no BOND for them to take money from. Government lawyers "Speak with forked tongue."

All government agencies, the COURTS, the prisons, the state troopers, the cities and even the States are run as non-profit corporations. Corporations are dead entities. Each corporation has to have at least three administrators such as a president, a Secretary and a Treasurer. They also have to have a Board of Directors. Most all COURTS are non-profit corporations run as a business to collect money from people. The most valuable thing you unwittingly have is the Bond they put on your head at birth. The bonds are worth millions.

Enough about this subject for now—if you want to know more read MAKE ALASKA GREAT AGAIN. I am simply trying to tell you a little about how the system works and how you are being manipulated your entire life.

REDUCE STRESS INFLAMATION, BE HAPPY AND LIVE LONGER

Stress is your #1 enemy

Histamines are part of your body's defense system. When you are emotionally stressed histamine mediates the release of hormones and other neurotransmitters as part of the stress response. A specific histamines receptor (H1R) is known to signal danger is present (free floating anxiety).

Histamine regulates the hormone leptin, which in turn regulates the feeling of hunger, improper function of this hormone can lead to obesity. Histamine levels affect insulin resistance, diabetes and cholesterol. Chewing your food will induce activation of histamine neurons, which in turn suppresses food intake through H1R activation.

Most people think stress as something that is caused by external environmental sources like overbearing boss, a nagging spouse, financial problems, etc. But we regularly experience internal physiological stress hormonal, enzymal, and chemical reactions inside our bodies that lead to physical stress. There are many kinds of stress. Stress can be cause by the wrong exercise or by eating unhealthy foods, by short days and long nights of winter. You need to recognize and manage it on multiple levels.

OXPHOS

One type of stress reaction is that it impedes oxidative phosphorylation (OXPHOS), the process by which cell use enzymes to oxidize nutrients to release the energy that is used to reform adenosine triphosphate (ATP).

ATP is considered to be the energy carrier of life. Plants also make and use ATP. It's the high energy molecule that stores the energy we need to do just about everything and it is the job of the mitochondria, the little alien cells within your human cells to produce ATP. OXPHOS releases 30 to 35 ATP molecules for any given source of energy.

The OXPHOS process also enables the re3lease of CO2 into the body, which is good for organs, tissues and oxygen absorption. Stress inhibits OXPHOS and slows down efficient respiration. Some people suffering too much stress may experience glycolysis, which is an inefficient form of respiration in comparison to OXPHOS, releasing only five to eight molecules of ATP. It also limits the amount of CO2 available to the body. **<u>Glycolysis can also be responsible for a "cancer metabolism."</u>**

Obviously, you don't want that. If your stress causes glycolysis, your cells will suffer as well as your overall health. Cancer metabolism goes downhill and causes metabolic slow and low energy, low temperature, low heart rate and the inability to fight disease and cancer. I cover yogi breathing exercise and the correct way to breathe. The breath of life is explained later in this book.

INTERNAL INFLAMATION = REAL STRESS

Internal inflammation is the other major symptom of stress. The easiest explanation is "leaky gut". This is when isn't preforming correctly. Everyone's gut contains bacteria. You need to ensure your gut has enough "good" bacteria as opposed to bad bacteria.

Leaky gut is when the bacteria get through your intestinal lining and attack other organs in your body. Many body functions you take for granted may shut down in people with leaky gut because the body goes into survival mode to fight off invasive bacteria.

When you don't have good health you may feel cold. Your hands may feel cold or your feet and nose. If you are feeling cold it's a sign of low metabolic function. It's a sign of cortisol metabolism. Another sign of low metabolism is a low heart rate. Your heart should be beating around seventy or eighty beats per minute.

When you have forty or fifty beats per minute it indicated you have low metabolism. This can cause a lot of problems. Not getting enough oxygen and nutrients to the tissues and organs. Do you experience chronic thirst? How often do you have to get up during the night to pee? If you have to get up more than four times a night to pee it's a sign of low metabolism.

All these are signs of gut inflammation and internal inflammation. Other reasons m be prostate pressing against the bladder, preventing it from emptying completely. Often these symptoms are a result of gut inflammation.

The medical profession is only now beginning to acknowledge that the leaky gut problem exists yet the intestinal permeability has been discussed in the medical literature for over 100 years. Yet today's doctors aren't even aware of it. The father of medicine. Hippocrates said "All disease begins in the gut."

How often do you have a bowel movement? Do you suffer from constipation or diarrhea? What is your stool like? Is it runny? Is it small pebbles? Does it have undigested food in it? The longer your food stays in the gut it rots. It ferments as it rots. However if the food passes through quickly there is less rotting. When you have a higher metabolism your food passes through your gut rapidly thus reducing the amount of harmful bacteria, yeast and other damaging things that our body has to fight.

ENDOTOXINS

It's not just the bacteria. All of them die and fight each other causing what are called endotoxins (bacteria fragments) that get into body that we have to continually fight off. Endotoxins contribute to inflammation. Raising your metabolism and exercising can move food through your gut faster. Having a bowel movement twice a day is a good thing. You will feel much more normal. You will not need laxatives.

SLEEP IS VERY INPORTANT

Are you restless and interrupted feeling exhausted when you wake up in the morning? So you wake up at night with a pounding heart? Do you wake up feeling refre3shed? If you are not feeling refreshed you are running on adrenaline and cortisol. A cortisol metabolism may mean you are low on thyroid hormones.

A good indicator of how good your mitochondria are doing their job is how fast your hair and nails are growing. Is your hair dry and brittle? When you get a cut, how fast does it heal? When your thyroid is not working properly a hormone called parathyroid slows healing and causes many problems in your body. You might be deficient in calcium or vitamin D3, vitamin K2. You might be getting too much phosphorus and not enough magnesium.

How is your sexuality? Do you care less if you ever have sex again? Is your libido low? How is your energy? Are you dragging around all day? Do you get sick often? All these symptoms are usually caused by gut inflammation. Eighty percent of your immune system is in the gut because of all the bacteria that are in there. When they are leaking into your other organs you get inflammation.

The main tasks your body's immune system are:

1. Neutralizing and removing pathogens like bacteria, virus, parasite and fungi.
2. Recognizing and neutralizing harmful substances that accumulates in your body from the environment.
3. Fighting against the body's own cells that have changed due to illness or cancer.

CLEAN WATER IS MOST IMPORTANT

Today most people drink tap water that contains chlorine to purify it and fluoride which is rat poison. Stanis fluoride is a toxic byproduct of oil refining. It was marketed as rat poison in the 1930's. Adolph Hitler fed it to prisoners to make them compliant and not try to escape. Long term exposure collects in the brain making it difficult for you to think.

Water traveling through circular pipes loses its electrical charge causing the molecules of H2O to be less wet. It has been proven that water also has a memory. Water processed and used several times over and over in the larger cities from sewage and agriculture runoff containing insecticide and memory of traveling tough the digestive systems of cattle and insects may alter your thinking in such a way that you may be diagnosed with mental disorders. Charged rain water with two hydrogen molecules at a wider angle can pass through capillaries and tissue more easily. Pure charged water is absolutely important for longevity.

The medical profession sponsored by the Rockefeller Foundation decided to endorse fluoride as tooth decay preventive. As I mentioned earlier the Rockefeller agenda is to produce obedient compliant workers with just enough intelligence to hold down a job but not to rise up above their station in life. They don't want competition.

The Rockefeller Foundation also picks the school curriculums so the kids who graduate can't even balance a check book. It's not their fault. It's our educational system. The teachers are hamstrung with the curriculums. The schools are designed to keep the children under constant stress with loud bells ringing every hour. Students are forced to shuffle down long hallways as fast as they can to get to the next class. If that isn't enough you have to remember the combination to your locker while being bullied by older and bigger students. Lunch consists of junk food to rot your brain.

At the present time the birth rate in the United States is only 1.6 children per couple. In order to maintain a population we need a birthrate of 2.3 to 2.6 children. The United States is losing its workforce. There will not be enough people workers to pay the Social Security Tax for all the baby boomers who are entering retirement. This is why the Democrats want thousands of immigrants to come into the United States across the southern border. As an added bonus once they are given work visas and a driver's license they vote Democrat.

China's workforce is in real trouble due to the Gates Foundation suggestion to adopt the one child per couple mandate for fear of overpopulation. Couples were forced to euphonize girls. Now China has a large population of angry young men without enough women to marry.

Fluoride does harden dental enamel but too much of it can darken and make the enamel crack. It may be the main cause of Alzheimer's in the elderly as it accumulates over many decades. Drinking dead water from plastic bottles is not good. Plastic accumulates in the blood and brain. Your body doesn't know what to do with it because it is a foreign substance so it stores it along with other toxins in fat tissue… My well is 100-feet deep and the snow melt if filtered through twenty miles of glacier silt and rock before it reaches our well. It also tastes good.

Where the water turned deadly:

Flouride system wasn't the only thing that failed in Hooper Bay
Author: **David Hulen**
Updated: July 1, 2016 Published September 22, 1992

This story was originally published on Sept. 22, 1992

Third in a series—From a letter signed by 300 Hooper Bay residents and sent last winter to Gov. Wally Hickel, asking for upgrades to the village's water and sewer systems at HOOPER BAY—Dominic Smith didn't realize that water from the village well was killing him. So he kept drinking.

The sicker he got, the more he drank. The more he drank, the sicker he got. All around his part of the village, his neighbors were falling sick, too.

By the next day, Smith was dead. His sister was in the hospital, critically ill. Sometimes water and sanitation systems in Alaska villages fail. Sometimes village governments fail, too. So do government regulators.

All those failures combined Memorial Day weekend in Hooper Bay.

A system that pumps a fluoride solution into the town's drinking water badly malfunctioned, poisoning a large portion of the population. Fluoride at levels 40 times what the federal government considers safe for deterring tooth decay were measured in drinking water, and epidemiologists later concluded that it was probably the most widespread fluoride poisoning ever documented, with more than 200 people estimated to have been stricken.

Jessie Ventura says the Nazis fed fluoride to prisoners in their concentration camps.

The first use of fluoride as a weapon was in World War ll Nazi concentration camps. Allegedly, the Nazis used it against the Jews to make them stupid, docile, and not have any will of their own by putting extremely high doses of fluoride in the water. In our water it is at a safe, optimized level, however fluoride in extreme doses can lead to these effects. German soldiers had little concern of the effects it had on teeth. They used fluoride to sterilize their prisoners as well as keep them calm and under control. This may be one reason there was not much uprising inside the camps.

Wikipedia: **"In the 1930`s, Hitler and the German Nazi's envisioned a world to be dominated and controlled by a Nazi philosophy of pan-Germanism.** The German chemists worked out a very ingenious and far-reaching plan of mass-control which was submitted to and adopted by the German General Staff. This plan was to control the population in any given area through mass medication of drinking water supplies. By this method they could control the population in whole areas, reduce population by water medication that would produce sterility in women, and so on. In this scheme of mass-control, sodium fluoride occupied a prominent place. ... «*Repeated doses of* **infinitesimal amounts** of fluoride will in time reduce an individual`s power to resist domination, by slowly poisoning and narcotizing a certain area of the brain, thus making him submissive to the will of those who wish to govern him.

[A convenient light lobotomy]

"The real reason behind water fluoridation is not to benefit children`s teeth. If this were the real reason there are many ways in which it could be done that are much easier, cheaper, and far more effective. The real purpose behind water fluoridation is to reduce the resistance of the masses to domination and control and loss of liberty.

«*When the Nazis under Hitler decided to go into Poland, both the German General Staff and the Russian General Staff exchanged scientific and military ideas, plans, and personnel, and the scheme of mass control through water medication* was seized upon by the Russian Communists because it fitted ideally into their plan to communize the world…

"I was told of this entire scheme by a German chemist who was an official of the great IG Farben chemical industries and was also prominent in the Nazi movement at the time. I say this with all the earnestness and sincerity of a scientist who has spent nearly 20 years' research into the chemistry, biochemistry, physiology and pathology of fluorine--any person who drinks artificially fluorinated water for a period of one year or more will never again be **the same person mentally or physically.**"

"Repeated doses of infinitesimal amounts of fluoride will in time reduce an individual's power to resist domination by slowly poisoning and narcotising a certain area of the brain and will thus make him submissive to the will of those who wish to govern him. Both the Germans and the Russians added sodium fluoride to the drinking water of prisoners of war to make them stupid and docile."

Charles Perkins (A research worker in chemistry, biochemistry, physiology) "History shows, actually, that in Nazi Germany, one of the first things that they did was add fluoride to the water in the ghettos where the Jews stayed" Matt Leffler

Believe it or not they put fluoride in baby water. They also put it in baby toothpaste. I went down to the local market and get a bottle of baby water and scanned the label for this book.

I discovered two brands of fluoridated baby water in local stores. Unbelievable! Would you let your baby drink fluoridated water? They actually sell fluoridated water for babies in stores. Mothers are so brainwashed that they think their baby needs fluoride.

Unbelievable! Would you let your baby use fluoridated toothpaste? There are serious topics you should act on immediately. Please buy fluoride free toothpaste. Toms brand that has one toothpaste that is fluoride free. If you can't find fluoride free toothpaste use baking soda. It's cheaper and it works to whiten your teeth. It also reduces acid in your mouth that can cause tooth decay.

CHAPTER FOUR

VACCINES

Would you inject mercury into your child? When I was a child we were injected with only two or three vaccines like small pox and tetanus. Ten years later they were injecting all of us with polio vaccines and others; can't remember. Now children are getting five to ten vaccines before age three.

There's big money in making and selling vaccines. Every few years they come up with a new one for something. Gardasil is currently all the rage. The government wants to inject your child with a vaccine that is supposed to ward off ovarian cancer. Some speculate that the vaccine causes sterility in children. Is this part of the "One World Government population reduction plan?"

Thiomersal (or Thimerosal) is a mercury compound used as a preservative used in some vaccines. Anti-vaccination activists promoting the incorrect claim that vaccination causes autism have asserted that the mercury in thiomersal is the cause. There is no scientific evidence to support this claim. Yea that's what big pharmacy wants you to believe. Autism rates are increasing at an alarming rate. Autism used to occur one in a hundred. Now one in ten children have some form of autism and it is probably due to the increasing amounts of mercury injected into their bodies.

How much mercury is in vaccines?

A vaccine containing 0.01% thimerosal as a preservative contains 50 micrograms of thimerosal per 0.5 mL dose or approximately 25 micrograms of mercury per 0.5 mL dose. For comparison, this is roughly the same amount of elemental mercury contained in a 3 ounce can of tuna fish.

That's a little misleading because when you eat a can of tuna the tuna is digested in the stomach and lower intestines and much of the mercury is expelled. When you inject it into the body directly there is no way for the body to eliminate it except to pass it through the liver and into the intestines to be excreted. Most of the mercury stays in the body. The mercury that passes through the blood brain barrier is stored in the fat in the brain and will impair brain function depending on the number of injections you give your child.

What vaccines still have mercury in them?

A few vaccines contained other preservatives, and they still do. Some other vaccines, including the measles, mumps, and rubella vaccine (MMR) never contained any preservative or any mercury.

When did Mercury stop being used in vaccines?

However, in July 1999, the Public Health Service agencies, the American Academy of Pediatrics, and vaccine manufacturers agreed that thimerosal should be reduced or eliminated in vaccines as a precautionary measure.

How much mercury is in a flu shot?

The call to remove it from vaccines in 1999 was prompted by the recognition that babies in their first 6 months could receive a cumulative dose of mercury exceeding federal guidelines. Injectable flu vaccine contains 25 mcg of thimerosal per 0.5-mL dose.

Health effects of mercury exposure

The inhalation of mercury vapor can produce harmful effects on the nervous, digestive and immune systems, lungs and kidneys, and may be fatal. The inorganic salts of mercury are corrosive to the skin, eyes and gastrointestinal tract, and may induce kidney toxicity if ingested. Mar 31, 2017

World Health Organization key facts:

Mercury is a naturally occurring element that is found in air, water and soil.

Exposure to mercury – even small amounts – may cause serious health problems, and is a threat to the development of the child in utero and early in life.

Mercury may have toxic effects on the nervous, digestive and immune systems, and on lungs, kidneys, skin and eyes.

Mercury is considered by WHO as one of the top ten chemicals or groups of chemicals of major public health concern. People are mainly exposed to methyl mercury, an organic compound, when they eat fish and shellfish that contain the compound.

Methyl mercury is very different to ethyl mercury. Ethyl mercury is used as a preservative in some vaccines and does not pose a health risk. *Not true!*

What is important to note here is that methyl mercury is a naturally occurring compound found in deep water fish that are high up on the food chain. It occurs in sea water in low concentrations. The microscopic plants that grow in oxygen rich sea water store a small amount of this natural compound. The krill and zoo or animal plankton eat the plant plankton further concentrating more amounts of methyl mercury.

The needle fish and herring eat the krill further increasing the concentration of methyl mercury. The tuna feed on the needle fish and herring have slightly more methyl mercury than salmon because salmon feed mostly on the krill. Then you have the swordfish that feed on the larger fish so obviously they will have a higher content of methyl mercury in their flesh. I believe you can eat as much of any fish from the ocean as you desire without affecting your brain and other organs because it is a natural occurring substance.

Ethyl mercury on the other hand is not a natural substance. It is not good to put in your body.

Mercury exists in various forms: elemental (or metallic) and inorganic (to which people may be exposed through their occupation); and organic (e.g., methyl mercury, to which people may be exposed through their diet). These forms of mercury differ in their degree of toxicity and in their effects on the nervous, digestive and immune systems, and on lungs, kidneys, skin and eyes.

Mercury occurs naturally in the earth's crust. It is released into the environment from volcanic activity, weathering of rocks and as a result of human activity. Human activity is the main cause of mercury releases, particularly coal-fired power stations, residential coal burning for heating and cooking, industrial processes, waste incinerators and as a result of mining for mercury, gold and other metals.

Once in the environment, mercury can be transformed by bacteria into methyl mercury. Methyl mercury then bio accumulates (bioaccumulation occurs when an organism contains higher concentrations of the substance than do the surroundings) in fish and shellfish. Methyl mercury also biomagnifies. For example, large predatory fish are more likely to have high levels of mercury as a result of eating many smaller fish that have acquired mercury through ingestion of plankton.

Generally, two groups are more sensitive to the effects of mercury. Fetuses are most susceptible to developmental effects due to mercury. Methyl mercury exposure in the womb can result from a mother's consumption of fish and shellfish. It can adversely affect a baby's growing brain and nervous system. The primary health effect of methyl mercury is impaired neurological development. Therefore, cognitive thinking, memory, attention, language, and fine motor and visual spatial skills may be affected in children who were exposed to methyl mercury as fetuses.

The second group is people who are regularly exposed (chronic exposure) to high levels of mercury (such as populations that rely on subsistence fishing or people who are occupationally exposed). Among selected subsistence fishing populations, between 1.5/1000 and 17/1000 children showed cognitive impairment (mild mental retardation) caused by the consumption of fish containing mercury. These included populations in Brazil, Canada, China, Columbia and Greenland.

A significant example of mercury exposure affecting public health occurred in Minamata, Japan, between 1932 and 1968, where a factory producing acetic acid discharged waste liquid into Minamata Bay. The discharge included high concentrations of methyl mercury. The bay was rich in fish and shellfish, providing the main livelihood for local residents and fishermen from other areas.

For many years, no one realized that the fish were contaminated with mercury, and that it was causing a strange disease in the local community and in other districts. At least 50 000 people were affected to some extent and more than 2000 cases of Minamata disease were certified. Minamata disease peaked in the 1950s, with severe cases suffering brain damage, paralysis, incoherent speech and delirium. However you have to realize that the people affected were eating fish every day.

Pandemic blows lid off laws limiting mercury in vaccines

Charlotte Schubert

Nature Medicine volume 16, page 9 (2010) Cite this article Published 2010.

FLU SHOT FACTS

Dr. Joseph Mercola

www.mercola.com

"If a person has had five consecutive flu shots – their chance of contracting Alzheimer's disease is 10x higher than the non-vaccinated person." By Dr. Hugh Fudenberg M.D., Ph.D. – author of over 80 research papers in Medicine, Biology and Immunology.

Vaccine Ingredients:
Thimerosal (Mercury Disinfectant/Preservative)
Aluminum (Additive to promote antibody response)
Formaldehyde (Disinfectant)
Ethylene Blycol (Anti-Freeze – deadly poison)
Phenol (Disinfectant, Dye)
Bensethonium Chloride (Antiseptic)
Methylparaben (Antifungal, preservative)

Mercury toxicity can result in brain injury and auto-immune injury. Aluminum is associated with Alzheimer's disease and seizures. Formaldehyde causes cancer and is considered hazardous waste that is no longer permitted as an ingredient in building insulation.

As public health officials around the world scramble to protect their citizens from swine flu, some in the US are grappling with an additional issue: state laws that limit the use of a mercury preservative in vaccines.

Between 2004 and 2006, six states—California, Delaware, Illinois, Missouri, New York and Washington—enacted laws limiting the use of the preservative, thimerosal, in flu vaccines and other shots given to children and pregnant women. The move was made in response to fears that thimerosal might cause neurological conditions such as autism.

This fall, however, all six states temporarily lifted the restriction in response to the outbreak of pandemic swine flu and the shortage of H1N1 vaccine. Many of the available vaccines came in multidose vials, which are quicker to manufacture and contain thimerosal to prevent contamination with repeated inserts of a needle.

Although the laws allow for such temporary suspensions, many people question their utility in the first place, given their feeble scientific foundation. The laws are "absolutely not" supported by research, says Diane Peterson, an associate director at the Immunization Action Coalition, a vaccine advocacy group in St. Paul, Minnesota. Even in 2004, before the laws were enacted, a US Institute of Medicine panel found that the scientific evidence did not support a link between thimerosal-containing vaccines and autism, and several subsequent studies have backed up that conclusion (N. Engl. J. Med. 357, 1281–1292; 2007)

But vaccine experts say that anti-thimerosal advocates remain vocal. Mary Selecky, Washington's secretary of health, received a flurry of complaints after she suspended the thimerosal restrictions in September. "They were quite angry at me," she says. "Clearly, folks were very pointed about feeling I was wrong."

Peterson is concerned that the thimerosal bans have helped fuel a backlash against vaccination. "The laws have contributed to the doubts people have about the safety of vaccines," she says. An ABC/Washington Post poll from October found that nearly half of parents did not intend to vaccinate their children against swine flu, in part because of safety concerns.

But as the scientific evidence mounts, laws limiting thimerosal have been harder to get through state legislatures. This year alone, advocacy groups and individuals in 12 states tried—and failed—to pass similar restrictions, according to Peterson. What's more, the laws are mostly redundant, as thimerosal was removed from nearly all vaccines by 2001. Even so, the six states' thimerosal bans remain on the books; Washington's is scheduled to go back into effect in March.

Why Democrats want millions of immigrants!

Obviously, bringing more people into the US increases the number of Democratic voters. It is an oxymoron in a way because Democrats are in favor of abortion. However I believe there are other reasons why they need people to migrate to the United States.

The birth rate is declining. In 2918 it was 1.73 births per woman. For the population in a given area to remain stable, an overall total fertility rate of 2.1 is needed, assuming no immigration or emigration occurs. That number dropped to 1.3 births per couple in 2020. Stewart said birth rates in the U.S. have slowly been declining for at least the past 50 years. The decline is the result of increasing levels of education and employment opportunities for women, and shifting gender roles, she said.—Mar 15, 2021 Blame it on woman's lib.

There is a big difference between 1.73 births and 2.1. In order for any given population to grow which we need to pay the Social Security for all the baby boomers we need a birth rate much higher than 2.1.

Winston Churchill was a great leader and the Prime Minister of England during World War Two. During those terrible times we lost a lot of good people. Winston Churchill said, "Couples should have four children; two to replace the mother and father, one to increase the population and one to offset mortality."

CHAPTER FIVE

BIG THREATS TO ALL LIFE ON EARTH

A 25-MILLION research study conducted by the national Toxicology Program (NTP) included two separate animal studies that reveled exposure to the type of radiation emitted by 2G and 3G cell phones could cause heart tumors, DNA damage and tumors in the brain, prostate, liver and pancreas. With 5G technology, these concerns are exponentially elevated. According to research by Dr. Yael Stein of Hebrew University of Jerusalem, computer simulations have revealed that seat gland concentrated MMWs in human skin, which may be sensed as heat waves or physical pain by humans. IN a 2016 letter to the Federal Communications Commission (FCC) in opposition to 5G technology, Stein wrote: "If these devices fill the public space they will affect everyone, especially the more susceptible members of the public – babies, pregnant women, the elderly, the sick and electro-hypersensitive individuals.

Potentially, if 5G WI FI is spread in the public domain we may expect more of the health effects currently seen with R$F/microwave frequencies, including many more cases of hypersensitivity, (EHS), as well as many new complaints of physical pain and a yet unknown variety of neurological disturbances. It will be possible to show a causal relationship between 5G technology and these specific health effects. The affected individuals may be eligible for compensation."

In addition a host of other research shows the proliferation of 5G for the sake of faster internet could be a public health disaster. For instance, the frequencies of 5G have also been linked to lens opacity in rats, which is linked to the formation of cataracts, impacted heart rate variability and heart changes (arrhythmias) in frogs; and immune system effects in healthy mice.

There are a serious environmental issues to consider as well, because MMWs are absorbed by both plants and rains. Studies have already shown that MMWs may invoke stress protein changes in plants such as wheat shoots, while low levels of non-ionizing radiation have been linked to disturbances and health problems in birds and bees. It could even pose a danger to the food supply via its potential absorption by plants. Humans and animals alike consume plant as a food source. The effects 5G has on plants could leave us with food that's not safe to eat. Think GMO's on steroids!

Many are not aware that the global 5G and 9G wireless network requires the use of rockets to deploy tens of thousands of satellites whizzing along at 28,000 miles only fifty miles above us. Several nations including Russia, China and the United States are each sending up many thousands of low earth orbiting satellites all emitting lots of EMF only fifty miles above our heads.

Once installed in your neighborhood, you won't have a choice to opt out of 5G exposure. An open letter to medical organizations by the Global Union Against Radiation Deployment From Space (GUARDS), points out, "Flooding the planet with microwave radiation in this way is a violation of human rights.

Space based microwave radiation deployments threated to inundate the planet with RF radiation without informed individual consent or a meaningful option for individual avoidance. Pulse-modulation microwave radiation has been reported by thousands of peer-review studies to cause cancer, DNA damage, infertility and blood-brain barrier damage. It impairs immune, endocrine, cardiovascular, and neurological functions in humans, and ultimately adversely affects all living things. Some satellites will use a very hi-frequency portion of the spectrum MMWs) shown to cause cellular resonance effects."

I you want to get involved in keeping 5G or the higher frequency 6g, 7G, 8G and 9G out of your community, the country and the world, please contact your legislators to voice your opposition. Parents for Safe Technology had more information, and a list of U.S. agencies you can contact about this important issue for the health and safety of future generations and the planet.

The alternative if you want to avoid cancer and other health issues you must live in a remote area fifty miles from the nearest cell tower in and underground bunker or cave shielded with metal roofing all around.

The second scariest threat to life on the planet

The hundreds of bio labs all over the world experiments with gene splicing and modification of DNA in bacteria and virus. A couple years ago the corona virus was released from a lab in China. That isn't the worst of it. That one tiny virus mutated several times into a more infectious and deadly pathogen.

Last year a lab in Texas modified a bacterium that can turn crop waste into alcohol. They were thinking the alcohol could be used to power farm machinery. Fortunately an environmentalist watchdog group shut them down and destroyed the bacteria. If it had gotten loose into the atmosphere and circular the planet every living thing would be turned into alcohol. All life as we know it including humans would be gone! Maybe we should shut down the bio labs and stop the irresponsible idiots trying to make a buck with gene modification. It's the money that turns everything to shit.

RADON GAS

If your house is built on solid granite rock the decaying radium in the granite continually releases a radioactive radon gas. Breathing this gas over time can cause cancer.

Radon is a radioactive gas that has been found in homes all over the United States. It comes from the natural breakdown of uranium in soil, rock, and water and gets into the air you breathe. Radon typically moves up through the ground to the air above and into your home through cracks and other holes in the foundation.

What are the symptoms of radon in your home?

A persistent cough could be a sign that you have radon poisoning.

Persistent cough.

Hoarseness.

Wheezing.

Shortness of breath.

Coughing up blood.

Chest pain.

Frequent infections like bronchitis and pneumonia.

Loss of appetite.

How do you get rid of radon in your home?

In some cases, radon levels can be lowered by ventilating the crawlspace passively (without the use of a fan) or actively (with the use of a fan). Crawlspace ventilation may lower indoor radon levels both by reducing the home's suction on the soil and by diluting the radon beneath the house.

Radon is the single largest source of radiation for almost everyone in Washington. Radon is the leading cause of lung cancer among nonsmokers, according to the U.S. Environmental Protection Agency, and the second leading cause behind smoking.

What part of the United States has a problem with radon?

Radon is present everywhere in the United States. Levels of the gas differ from state to state, but it is particularly high in North Dakota and Iowa. The U.S. Environmental Protection Agency's Map of Radon Zones shows the potential for elevated radon levels for each county in the United States. —Sep 14, 2015

Earth is a giant atomic pile with all kinds of nasty radioactive elements at its core. Thorium ranks 31st in abundance in Earth's crust. It has a half-life of fifty billion years. This contradicts the 'Big Bang Theory' that everything in the universe is only 14-billion years old. Scientists don't know what happens to photos after traveling 14-billion years. They think galaxies are traveling away from us at the speed of light thus confirming a Catholic Monk's big bang explosion creation theory. After traveling 14-billion years at the speed of light photons might actually loose energy or wear out shifting to the red. Nobody has been around long enough to prove this.

CHAPTER SIX

THE CORONAVIRUS SIMULATION

Live Simulation Exercise to Prepare Public and Private Leaders for Pandemic Response

Published 15 Oct 2019

Kirsten Salyer, Public Engagement, Tel.: +41 79 265 8773; Email: kirsten.salyer@weforum.org

- The world has seen a growing number of epidemic events, amounting to about 200 per year
- Pandemics could cause average annual economic losses of 0.7% of global GDP
- Event 201 exercise will bring together public and private leaders to inform multistakeholder cooperation for pandemic preparedness and response
- Follow the live virtual exercise from 08.50 - 12.30 EDT here

Geneva, Switzerland, 15 October 2019 – The Johns Hopkins Center for Health Security in partnership with the World Economic Forum and the Bill & Melinda Gates Foundation will host Event 201: a high-level simulation exercise for pandemic preparedness and response, in New York, USA, on Friday 18 October, 08.45 - 12.30 EDT.

The exercise will bring together business, government, security and public health leaders to address a hypothetical global pandemic scenario. It will also feature a live virtual experience from 08.50 – 12.30 EDT to engage stakeholders worldwide and members of the public in a meaningful conversation of difficult high-level policy choices that could arise in the midst of a severe pandemic.

Look up Codex Alamentarious. Don't you find it odd that the World Economic Forum with Bill Gates present would hold meetings prior to releasing the coronavirus?

The world has seen a growing number of epidemics in recent years, with about 200 events annually including Ebola, Zika, MERS and SARS. At the same time, collective vulnerability to the social and economic impacts of infectious disease crises appears to be increasing. Experts suggest there is a growing likelihood of one of these events becoming a global threat – or an "event 201" pandemic – that would pose disruptions to health and society and causes average annual economic losses of 0.7% global GDP, similar in scale to climate change.

"We are in a new era of epidemic risk, where essential public-private cooperation remains challenged, despite being necessary to mitigate risk and impact" said Arnaud Bernaert, Head of Shaping the Future of Health and Health Care, World Economic Forum. "Now is the time to scale up cooperation between national governments, key international institutions and critical industries, to enhance global capacity for preparedness and response."

The International Health Regulations (IHR) that unite 196 countries across the globe in a legal commitment to prevent and respond to acute public health risks, prioritize both minimizing public health risks and avoiding unnecessary interference with international traffic and trade. Minimizing the economic impact of epidemics also represents an opportunity to build core capacities to prevent, detect, and respond to outbreaks generally.

"We live in an increasingly interconnected world, and we must help all UN member states align with the International Health Regulations and be prepared to prevent, detect, and respond to acute outbreaks," said Chris Elias, President of Global Development at the Gates Foundation. "If we fail to do so, the world will be unprepared for the next pandemic."

"In this new era of extreme pandemic threat, public-private cooperation is essential for an effective response," said Tom Inglesby, Director of the

Johns Hopkins Center for Health Security. "While governments and public health systems are already strained due to the increase in dangerous outbreaks, experts agree that a severe, fast-spreading human-to-human pandemic incident could happen at any time. We believe this well-crafted and thorough realistic tabletop exercise will provide leaders with a deeper understanding of the impact of epidemics on their communities and inspire them to take important steps to advance prevention and response."

The participants in the live simulation represent a range of backgrounds and industries and include:

- Latoya Abbott, Risk Management/Global Senior Director Occupational Health Services, Marriott International
- Stan Bergman, Chairman and CEO, Henry Schein
- Sofia Borges, Senior Vice President, UN Foundation
- Chris Elias, President, Global Development division, Bill & Melinda Gates Foundation
- Tim Evans, Former Senior Director of Health, World Bank Group
- George Gao, Director-General, Chinese Center for Disease Control
- Avril Haines, Former Deputy Director, Central Intelligence Agency; Former Deputy National Security Advisor
- Jane Halton, Board member, ANZ Bank; Former Secretary of Finance and Former Secretary of Health, Australia
- Matthew Harrington, Global President and Chief Operations Officer, Edelman
- Chikwe Ihekweazu, Director General, Nigeria Centre for Disease Control
- Martin Knuchel, Head of Crisis, Emergency and Business Continuity Management, Lufthansa Group Airlines
- Eduardo Martinez, President, The UPS Foundation
- Stephen Redd, Deputy Director for Public Health Service and Implementation Science, US CDC
- Paul Stoffels, M.D., Vice Chair of the Executive Committee and Chief Scientific Officer, Johnson & Johnson
- Hasti Taghi, Vice President and Executive Advisor, NBCUniversal Media
- Lavan Thiru, Chief Representative, Monetary Authority of Singapore

Similar high-level pandemic exercises designed to address difficult policy issues have included: <u>Dark Winter</u>, examining the challenges of a biological attack on the US; <u>Atlantic Storm</u>, asking NATO leaders to respond collaboratively to a bioterrorist attack: and most recently, <u>Clade X</u>, calling on US government leaders to make difficult national security and public health decisions in the face of a rapidly evolving global crisis.

In addition, Bill Gates co-chaired a simulation at the Forum's Annual Meeting 2017, resulting in the creation of the Epidemics Readiness Accelerator, a public-private platform to address effective readiness in issues including travel and tourism, supply chain and logistics, legal and regulatory, communications and data innovations.

Notes to editors
Learn more about the exercise <u>here</u> and follow the exercise at #Event201 and @JHSPH_CHS, @wef and @gatesfoundation

Join the live virtual experience and register at http://www.centerforhealth security.org/event201/

Learn more about the Forum's Epidemics Readiness Accelerator <u>here</u>

Read the Forum's report "Outbreak Readiness and Business Impact" <u>here</u>

Read about the Transformation of Biological Risk in the Forum's 2019 Global Risk Report <u>here</u>

Learn about the World Economic Forum's impact: <u>https://www.weforum. org/our-impact</u>

View the best Forum Flickr photos at <u>http://wef.ch/pix</u>

Become a fan of the Forum on Facebook at <u>http://wef.ch/facebook</u>

Follow the Forum on Twitter at <u>http://wef.ch/twitter</u>

Read the Forum blog at <u>http://wef.ch/agenda</u>

View upcoming Forum events at <u>http://wef.ch/events</u>

Subscribe to Forum news releases at http://wef.ch/news

The World Economic Forum, committed to improving the state of the world, is the International Organization for Public-Private Cooperation. The Forum engages the foremost political, business and other leaders of society to shape global, regional and industry agendas.

It has been my experience that most all government agencies do the opposite of what they portend to do. For example the World Economic Forum held a meeting in New York December of 2019 where they showed a fictional video of what would happen if the carina virus was released in South America. The fictional outcome was that it would kill 65,000 million people.

Share Event 201

The Johns Hopkins Center for Health Security in partnership with the World Economic Forum and the Bill and Melinda Gates Foundation hosted Event 201, a high-level pandemic exercise on October 18, 2019, in New York, NY. The exercise illustrated areas where public/private partnerships will be necessary during the response to a severe pandemic in order to diminish large-scale economic and societal consequences. Statement about nCoV and our pandemic exercise

Statement about nCoV and our pandemic exercise

In October 2019, the Johns Hopkins Center for Health Security hosted a pandemic tabletop exercise called Event 201 with partners, the World Economic Forum and the Bill & Melinda Gates Foundation. Recently, the Center for Health Security has received questions about whether that pandemic exercise predicted the current novel coronavirus outbreak in China.

To be clear, the Center for Health Security and partners did not make a prediction during our tabletop exercise. For the scenario, we modeled a fictional coronavirus pandemic, but we explicitly stated that it was not a prediction. Instead, the exercise served to highlight preparedness and response challenges that would likely arise in a very severe pandemic. We are not now predicting that the nCoV-2019 outbreak will kill 65 million people. Although our tabletop exercise included a mock novel coronavirus, the inputs we used for modeling the potential impact of that fictional virus are not similar to nCoV-2019.

Keep in mind that this was all done back in October 2019 about the same time the virus was being leaked to the 10,000 or so Chinese Wohan Provence protestors protesting air pollution.

The coronavirus 'pandemic' sure does have a lot of strange coincidence. What are the chances that Bill gates, The CDC, Big Pharma reps, the WEF and others unknowingly planned a "fictional" simulation for a coronavirus outbreak 2 months before an actual outbreak occurred? A big par of it was controlling the narrative afterwards, in including censorship of anti-vaccine voices. This is yet more evidence that the coronavirus was a planned event, not a random disease that happened to crows species at a seafood market.

The inventor of the PCR test Kary Mullis warned it could not be used to determine infectious etiology, i. e. the cause of a disease, since it is a manufacturing technique not a diagnostic tool. PCR amplifies exponentially, so any initial error or contaminates are amplified to create a false positive. The supposed nique viral sequence found in SARS-COVID-2 contains human DNA! The COVID PCR test is testing for something that exists in every human. The Tanzanian president sent in samples from a bird, goat and PAWPAW – and they al came back positive. The basis for declaring you positive and infectious and destroying you rights and livelihood, is a giant scam.

The Italian parliament calls for the arrest of Bill Gates for Crimes against Humanity.

The alleged pandemic is an excuse to inject all the people of the world with a substance that modifies your DNA for the purpose of shortening your lifespan. Bill Gates machinations and shenanigans are being noticed by people all around the world. In 2006 he gave 5-million dollars to a British corporation, the Pirbright foundation to manufacture a

In 2016 **Bill & Melinda Gates Foundation funds development of Pirbright's Livestock Antibody Hub supporting animal and human health**

Posted: 15 November, 2019

Researchers from The Pirbright Institute have been awarded US $5.5 million by the <u>Bill & Melinda Gates Foundation</u> to establish a Livestock Antibody Hub aimed at improving animal and human health globally. The ambitious program of work will see extensive collaboration between multiple UK research organizations in order to utilize research outcomes in livestock disease and immunology to support human health as part of the 'One Health' agenda.

Six leading scientists from Pirbright will be involved in the project, including <u>Professor John Hammond</u>, <u>Professor Venugopal Nair</u>, <u>Dr Simon Graham</u>, <u>Dr Elma Tchilian</u>, <u>Professor Munir Iqbal</u> and <u>Dr. Erica Bickerton</u>. Their combined expert knowledge will drive the study of cattle, pig and poultry antibody responses at high resolution to expand our understanding of protective immunity in species that can also be used as models for a range of human infectious diseases.

The aim is to use Pirbright's expertise in livestock viral diseases, cutting-edge technology and unique high-containment facilities to bring antibody discovery, manipulation and testing up to the benchmark already seen in the immunological field for rodents and humans. "New tools have given us the opportunity to utilize these detailed antibody responses to make the next generation of vaccines and therapies" said research lead Professor Hammond.

This highly collaborative work will address the needs of the livestock research community whilst bridging the requirements of the vaccine industry. A number of work programs will focus on studying B cells and antibodies at multiple scales including gene expression, single cell function and the entire antibody response.

Findings from this research will be used to drive vaccine selection and design and test antibody therapies, "which will improve animal health and ultimately human health, as well as ensuring the security of our food supply", finished Professor Hammond. Pirbright will ultimately act as a 'Hub' able to provide specific methods, access to animal models and the associated expertise to drive antibody research within the 'One Health' agenda.

"This is the single biggest investment in the immunology of livestock in the UK from an international funder, and the British Society for Immunology will do all we can to support this collaborative initiative and help maximize its impact for the benefit of human and animal health", commented Dr Doug Brown, Chief Executive of the <u>British Society for Immunology</u>.

The Bill & Melinda Gates Foundation

Gates Foundation pledged US$ 750 million to set up Gavi in 1999.

The Foundation is a key Gavi partner in vaccine market shaping. Bill and Melinda Gates' realization in the late 1990s that rotavirus was killing half a million children every year was one of the events that led them to set up their <u>Foundation</u> in 1994.

Infowars reported January 20, 2019 that the Pirbright Foundation filed for a patent on the coronavirus in 2016 and that they received the patent in 2018. After they received the patent Bill Gates gave 750-million for more research supposedly to help children—this from a person who is into eugenics and world population reduction. My opinion on the matters is once they had the patent they needed a vaccine for the rich elite one-world-government freaks they can sit back and watch people die.

Bill Gates, founder of Microsoft and the main orchestrator of Operation Coronavirus, is at it again. His old computer firm (where he still retains influence) has filed a patent (ending 060606 which is very close to 666) which involves a human implantable device for buying and selling crypto currency. Is this akin to the biblical Mark of the Beast? Guess what the name of the enzyme is for Gates quantum dot tattoo vaccination? Luciferase. No Kidding! This ties into immunity passports, mandatory vaccination ID2020 and Gates plan for every single person on Earth to have a digital ID), and the human micro chipping agenda. You just can't make this stuff up.

CHAPTER SEVEN

BIGGEST THREAT TO FREEDOM

America is the last bastion of freedom on earth.
If we lose it here the entire world will be enslaved.
"Resistance is futile!" "You will be assimilated!"
–Startreck Borg Collective

CRITICAL RACE THEORY; WHAT IT IS, HOW TO FIGHT IT.

By: Christopher F. Rufo, founder and director of Battlefront, a public policy research center. He is a graduate of Georgetown University and former Lincoln fellow at the Claremont Institute for the Study of Statesmanship and political Philosophy.

Critical Race theory is a threat to your very existence and the way you live. I may be getting a little off track here but in order for the human race to survive we need to become aware of changes that affect our freedom and our future. The following is adapted from a lecture delivered at Hillsdale College on March 30, 2021. Mankind cannot progress under slavery.

"Critical race theory is fast becoming America's new institutional orthodoxy. Yet most Americans have never heard of it—and of those who have, many don't understand it. It's time for this to change. We need to know what it is and how to fight it.

WHAT IT IS

Critical race theory is an academic discipline, formulated in the 1990s, built on the intellectual framework of identity based Marxism. Regulated for many years in universities and obscure academic journals, over the past decade it has increasingly become the default ideology in our public institutions. It has been injected into government agencies, public school systems, teacher training programs, human resources departments in the form of diversity training programs, human resource modules, public policy frameworks and school curricula.

There are a series of euphemisms deployed by its supporters to describe critical race theory, including "equity," "Social justice," diversity and inclusion," and "culturally responsive teachings." Critical race theorists, masters of language construction, realized that "neo Marxism" would be a hard sell. *Equity,* on the other hand, sounds non-threatening and is easily confused with the American principals of *equality.* But the distinction is vast and important. Indeed, equality—the principle proclaimed in the Declaration of Independence, defended in the civil War, and codified into law with the 14th and 15th Amendments. The civil rights Act of 1964, and the voting Rights Act of 1965—is explicitly rejected by critical race theorists. To them, equality represents "mere nondiscrimination" and provides "camouflage" for white supremacy, patriarchy, and oppression.

In contrast to equality, equity as defined and promoted by critical race theorists is little more than reformulated Marxism. In the name of equity, **UCLA Law Professors and critical race theorist Cheryl Harris has proposed suspending private property rights, seizing land and wealth and redistributing them along racial lines.**

Critical race guru Ibram X. Kendi, who directs the Center for Antiracist Research at boston University, has proposed to the creation of a federal Department of antiracism. This department would be independent of (i.e., unaccountable to) the elected branches of government, and would have the power to abolish any law at any level of government and curtail the speech of political leaders and other who are deemed insufficiently antiracist."

One practical result of the creation of such a department would be the overthrow of capitalism, since according to Kendi, "In order to truly be anti-capitalist." In other words, identity is the means and Marxism is the end.

An equity-based form of government would mean the end not only of private property, but also of individual rights, equality under the law, federalism, and freedom of speech. These would be replaced by race-based redistribution of wealth, group-based rights, active discrimination, and omnipotent bureaucratic authority. Historically, the accusation of "anti-Americanism" has been overused. But in this case, Its not a matter of interpretation—critical race theory prescribes a revolutionary program that would over-turn the principles of the Declaration and destroy the remaining structure of the Constitution.

HOW IT WORKS

What does critical race theory look like in practice? Last year, I authorized a series of reports focused on critical race theory in the federal government. The FBI was holding workshops on intersectional-ity theory. The Department of Homeland Security was telling white employees they were committing "micro inequities" and had been "socialized into oppressor roles."

The Treasury Department held training sessions telling staff members that "virtually all white people contribute to racism" and that they must convert "everyone in the federal government" to the ideology of "antiracism." And the Sandia National Laboratories, which designs America's nuclear arsenal, sent white male-executives to a three-day re-education camp, where they were told that "white male culture" was analogous to the "KKK," "white supremacists," and "mass killings."

The executives where then forced to renounce their "white male privilege" and write letters of apology to fictitious women and people of color."

This year, I produced another series of reports focused on critical race theory in education. I Cupertion, California, an elementary school force first graders to deconstruct their racial and sexual identities, and rank themselves according to the "power and privilege."

In Springfield, Missouri, a middle school forced teachers to locate themselves on "oppressor matrix," based on the idea that straight white,

English-speaking Christian males are members of the oppressor class and must atone for their privilege and "covert white supremacy."

In Philadelphia, an elementary school forced fifth-graders to celebrate "Black communism" and simulate a Black power rally to free 1060's radical Angela Davis from prison, where she had once been held on charges of murder.

And in Seattle, the school district told white teachers that they are guilty of "spirit murder" against black children and must "bankrupt [their] privilege to acknowledgement of [their] thieved inheritance."

I'm just one investigative journalist, but I've developed a database of more than 1,000 of these stories. When I say that critical race theory is becoming the operating ideology of our public institutions, it is not an exaggeration—from the universities to bureaucracies to k-12 school systems, critical race theory has permeated the collective intelligence and decision making process of American government, with no sign of slowing down.

This is a revolutionary change. When originally established, these government institutions are presented as neutral, technocratic, and oriented towards broadly-held perceptions of the public good. Today under the increasing sway of critical race theory and related ideologies, they are being turned against the American people. This isn't limited to the permanent bureaucracy in Washington, D. C., but is true even in red states, and is spreading to county public health departments, small Midwestern school districts, and more. This ideology will not stop until it has devoured all of our institutions.

FUTILE RESISTANCE

Thus far, attempts to halt the encroachment of critical race theory have been ineffective. There are a number of reasons for this. First, too many Americans have developed an acute fear of speaking up about social and political issues, especially those involving race. According to a recent Gallop poll, 77 percent of conservatives are afraid to share their political beliefs publicly. Worried about getting mobbed on social media, fired from their jobs, or worse, they remain quite, largely ceding the public debate to those pushing these anti-American ideologies. Consequently, the institutions themselves become monocultures: dogmatic, suspicious, and hostile to a diversity of opinion. Consequently the institutions themselves become monocultures: dogmatic, suspicious, and hostile to a diversity of opinion. Conservatives in both the federal government and public school systems have told me that their "equity and inclusion" departments serve as political offices. Searching for and stamping out any dissent from the official orthodoxy.

Second, critical race theorists have constructed their argument like a mouse trap/. Disagreement with their program becomes irrefutable evidence of a dissenter's "white fragility," "unconscious bias," or "internalized white supremacy." I've seen this projection of false consciousness on their opponents play out dozens of times in my reporting. Diversity trainers will make an outrageous claim—such as "all whites are intrinsically oppressors" or "oppressors and "oppressed"?

Do they support mandatory curricula teaching that "all white people play a part in perpetuating systemic racism"? Do they support public schools instructing white parents to become "white traitors" and advocate for "white abolition"? Do they want those who work in government to be required to undergo this king of reeducation? How about managers and worker in corporate America? How about the men and women in our military? How about every one of us?

These are all part of a successful strategy to defeat the forces of critical race theory, "government action, grass roots mobilization, and an appeal to principle."

We all see examples of government action. Last year, one of my reports led President Trump to issue an executive order banning critical race theory-based on training programs in the federal government. President Biden rescinded this order on his first day in office, bit it provides a model of government and municipal leaders to follow.

This year, several sates legislatures have introduced bills to achieve the same goal: prevents public institutions from conducting programs that stereotype scapegoat, or demean people on the basis of race. And I have organized a coalition of attorneys to file lawsuits against schools and governments agencies that impose critical race theory-based programs on grounds of the First Amendment (which protects citizens from compelled speech), the fourteenth Amendment (which provides equal protection under the law), and the Civil Rights Act of 1964 (which prohibits public institutions from discrimination on the basis of race).

On the grassroots level, a multiracial bipartisan coalition is emerging to do battle against critical race theory. Parents are mobilizing against racially divisive curricula in public schools and employees are increasingly speaking out against Orwellian reeducation in the workplace. When they see what is happening, Americans are naturally outraged that critical race theory promotes three ideas—race essentialism, collective guilt, and neo-=segregation—which violates the basic principles of equality and justice. Anecdotally, many Chinese-Americans have told me that having survived the Cultural Revolution in their former country, they refuse to let the same thing happen here.

In terms of principle, we need to employ our own moral language rather than allow ourselves to be confined by the categories of critical race theory. For example, we often find ourselves debating "diversity." Diversity as most of us understand is generally good, all things beings equal, but it is of secondary value. We should be talking about and aiming at excellence, a common standard that challenges people of all backgrounds to achieve their potential. On the scale of desirable ends, excellence beats diversity every time.

Similarly, in addition to pointing out the dishonesty of the historical narrative on which critical race theory is predicated, we must promote the true story of America history, but that places them in the context of our nation's high ideals and the progress we have made towards realizing them. Genuine American history is rich with stories of achievements and sacrifices that will move the hearts of Americans—in stark contrast to the grim and pessimistic narrative pressed by critical race theorists.

Above all, we must have courage—the fundamental virtue required in our time. Courage to stand and speak the truth. Courage to withstand epithets. Courage to face the mob. Courage to shrug off the scorn of the elites. When enough of us overcome the fear that currently prevents so many from speaking out, the hold of critical race theory will begin to slip away. And courage begets courage. It's easy to stop a line dissenter: it's much harder to stop 10, 20, 100, 1000, 100,000,000, or more who stand up together for the principles of America.

Truth and justice are on our side. If we can muster the courage, we will win.

CHAPTER EIGHT

HOW THE GODS CREATED MAN

At the present time sheep and cattle are being cloned in England, China and other countries around the world. Our science is slowly catching up with the ancient Anunnaki. There are over 100,000 ancient clay tablets in a London Museum that haven't been translated yet so we don't know the complete story of the origin of the human race.

The secret use of monatomic gold and platinum has been known by secret societies and kept from the common man for thousands of years. Only the rich and powerful were allowed to have this knowledge.

President Clinton, George Bush Junior, Senior and Al Gore all went to Yale University and all were members of the same fraternity, the Scull and Bones fraternity. This fraternity participated in rites involving the consumption of human bones. The scull in particular was prized above all else for the illuminating substance it contains.

Geronamo's bones were dug up and stolen from the Apache Indian Reservation by Yale fraternity brothers. The Apache Nation sued Yale University to get them back but was unsuccessful. The case has been dragging on for years. I bet they will never see those bones again. Clinton and the Bushes ate them.

[Readers note: For the last twenty-five years the United States has been ruled by cannibal presidents groomed by this secret society formed in Nazi Germany.] Monatomic minerals go through the blood-brain barrier and combine with other substances creating large molecules which are trapped in the brain. Unable to leave, they eventually migrate into the scull bones and collect there. The effect on a person is they get smarter. This is due to the fact that the brain super-conducts electricity more efficiently. A person with enough of this material in his or her brain can access the other Internet, which is a body of information that holds all knowledge, commonly known as the collective unconscious. This is the inter-dimensional plane that Edger Casey and other adepts were able to access. While in a trance Edger could answer any question pertaining to any subject.

Enki and his brother Enlil were sent to Earth 300,000 years ago to mine gold. Enki came first in a spaceship filled with water in a voyage that lasted eight years. This make sense because the water would protect him from space radiation and micro-meteorites traveling faster than a speeding bullet. Since space is very cold in the order of 250 degrees below zero Fahrenheit. Once a small meteorite pierced the hull of the ship the incompressible water would slow it to a stop and as the water leaked out though the hull it would freeze and plug the hole.

King Anu had sent Enki on this one-way, death-defying, voyage to Earth because he had gotten into some kind of trouble back on their planet Niberu. The Anunnaki weren't sure if he could survive the trip. Once he arrived on Earth he radioed back using the ARK transmitter and the white powder-gold material saying, 'that he was okay.' This white-powder-gold or ORMUS material exists in multiple space times transcending space and time so that using their advanced technology they could send message instantly to any part of the Galaxy without waiting hundreds of years for an answer radio communication traveling at the speed of light.

They take the hard metal and turned it into the antigravity white power-gold with electricity. In that state the orbitally rearranged atoms of gold and platinum don't attract to gravity so it was easy to take it off planet to their home planet Niberu. They spray the ORME gold into their upper atmosphere of their planet to stop cosmic radiation from ageing their populations and to revitalize everything growing on their planet. ORME gold allows plants as well as animals to grow better and faster. Orme gold can be transmuted into any element needed for compete cell division so that everyone living there can live hundreds or even thousands of years.

For several years Enki and his brother Enlil they tried taking gold out of sea water but the process didn't produce enough to suit their needs. They searched the entire earth for the largest deposit of gold eventually settling on South Africa. We are still mining gold there today.

They needed smaller mine workers to tunnel underground because the Anunnaki stood well over ten feet tall so they brought thousands of the much smaller Igagi to earth. The mine work went on for the next 150,000 or so years until the Igagi decided to go on strike. Enlil was a hard task master. Sumerian clay tablets say he killed and starving 17,000 of the much smaller Igagi until there were so few left that the gold production was falling behind.

They had to do something to increase production so they decided to use some of their own DNA to create the much smaller perfect modern man. Mauri or (Mary) the sister of Enki donated some of her eggs. It took may tries to clone the perfect man that could reproduce. He looked much like Pierce Brasman. They inserted indigenous hominid DNA into Mauri's eggs hundreds of times. They were seeking a man that would be smart enough to operate a pick and shovel and obey orders. They didn't want him to overpopulate the earth so the tweaked the DNA so he would not liv longer than 120 years. They needed a compatible female with similar blood type so they took a piece of Adam's rib bone DNA and planted it into one of Mauri's egg. Eventually Eve was cloned and able to mate with Adam. Viola they had obedient slave workers.

We all know the Biblical story about the two sons of Adam, Cane and Abel. They must have taken wives from the Neanderthal tribes in the area because we have no record of where there other humans came from. Evidently the new modern weaker humans had to have done some inbreeding and mated with other indigenous peoples.

The actual stone relief carving above shows one of the bird-like gods carrying a bucket of water and a pine cone. Maybe it isn't a pine cone. Maybe it is some kind of weapon resembling the pineal gland or a transmitter to control or broadcast into the minds of their enemies. What is that thing with rivets on the back of the other guy? Note the wrist watch on the arm of the guy holding the pine cone! There are also tubes running up from his wrist on his right arm holding the water bucket. Obviously they had technology beyond our comprehension.

Is that a watch on his wrist? Did they have 11 hour days? Is he activating the antenna? Why are there seven petals on the flowers or antenna? Why the cellular structure of the device in his hand?

Analysis of the skeletal structure of the more massive Neanderthal bones indicated that they could lift as much as 2000 pounds or one ton. The Anunnaki wanted a workforce small enough to go underground yet stupid enough to believe their overseers were Gods and gullible enough to believe almost anything without doing any critical thinking. The reason Adam and Eve were kicked out of the Garden of Eden was because they started to have knowledge of good and evil. Eating the fruit was a metaphor. The satisfaction of figuring something out on your own and understanding truth is much like the pleasure of eating tasty fruit. The Bible mentions that the Nephilim gods or Anunnaki were concerned that their slave offspring would eventually become like them having knowledge not just about good and evil but about all things.

Moses God: A definite bird of prey who fed on meat—possibly the occasional human baby. Note the wrist watch. Maybe they are cell phones or some kind of communication devices? Did these advanced races perform genetic modification? Was it wrong? Who are we to judge?

About thirty-five-thousand-years ago Enki, or Thoth's first-born, son Marduk (Horus) was born on the planet Nibiru. We don't know what planet this name refers to however it could be the planet Mars or even

Venus, given the number of structures seen in NASA photographs of Mars and Venus either one is a possibility. Another possibility is that Nibiru is the tenth planet Zecharia Sitchin wrote about in his books.

I have a theory that there were several Garden of Eden clowning factories located several different places around the world cloning species of mankind that were adapted to their specific climate and local. Or they may have eventually adapted to their specific locality through natural selection evolution although certain genetic markers like the color of hair and the slant of the eyes etc. would tend to negate that theory. I also believe that in their rush to produce enough workers quite a few of them escaped to form tribes when men fathered fifty to a hundred offspring to war against each other and the Anunnaki themselves. This may be the reason why they created the flood but there were many other reasons mentioned in my book, **COSMOLOGICAL ICE AGES**.

Truth is subject to the eye of the beholder.

The people who wrote the ancient books of the Bible wrote down what they saw but they didn't fully understand what they were seeing because they lacked knowledge of advanced science and technology. In Genesis, the first book of the Bible is a very accurate description of the creation event from the prospective of people not familiar with space travel or cloning.

First of all the word Nephelim is a plural word. It was mistranslated from the ancient Hebrew into Greek or some time when the Greek was translated into English. It is defined in various texts as the fallen ones. Angels or fallen ones are usually depicted old paintings as having wings so obviously they had some kind of way of traveling through the atmosphere on earth and possibly the ability to come down from space.

There several of them at the time working to clone various life forms in the level four, laboratory called the Garden of Eden. They didn't actually fall down from space as that would have killed them. They obviously were space men who had advanced flying machines,. There were at least five of them and one woman called, Mauri (Mary).

According to translations of ancient Sumerian scrolls King Anu sent Enki, one of his sons from their home Planet Niberu to Earth to look for gold. they burn it into the white powder that has little or no mass allowing it to be transported off earth because it weighs practically nothing and in some cases when charged with frequency becomes antigravity.

The Nephilim or Anunnaki tried hundreds of times to clone the perfect man who could reproduce. They were using some of their own DNA and inserting into the eggs of various indigenous hominids species from earth. The cloning process is tricky. For example when you cross a donkey with a horse you get a mule. Mules can't reproduce.

The Nephilim were trying to create a smaller replica of themselves with enough intelligence to mine underground. Apparently they stood fifteen to twenty feet tall. It becomes impracticable to mine underground when you are twenty-feet tall as the tunnel has to be large requiring the removal of too much rock when following a small vein of gold bearing quartz. Why do you think most all churches and government building have large doorways twenty to thirty-feet tall?

After they had the perfect man who could reproduce they needed a compatible mate with similar DNA. This is the reason they took part of Adam's rib bone.

Finally, they had a semi-intelligent man who worshiped the Nephilim as their creator Gods. When Eve discovered the truth and took Adam out of the level four facility they escaped into the wilderness. This alarmed the Nephilim because Genesis mentions that man eating from the tree of knowledge of good and evil would lead to their downfall because they would eventually see that their creator Gods were no longer Gods but advanced life forms from another planet.

Another version of the story is after they had created five different races of people using various DNA from and the seeds of Adam. They needed to do this because to create a race of people you can't keep on marring your sister as this leads to horrible DNA abnormalities and birth defects that are carried down through the generations. English royalty is plagued with the blood disease hemophilia where the person can bleed to death from a small cut. We see these things in our royalty who marry their first cousins and other close family members to keep wealth and land holding within the family.

Eventually, the five Nephilim began to quarrel so they each took their group of slave humans and went their separate ways to different parts of the earth. They eventually colonized the whole earth and surveying it marking off various boundaries to claim their land and used their slave humans to fight their battles over land and possessions. This would explain why there are different looking types not races of humans living all over the earth. Technically speaking, the human race is one race because race is defined as the ability to reproduce. As far as we know all humans can interbreed so basically, we are all one race.

The humans living in what is now known as England were starving as all they had to eat was wheat and barley. They didn't have potatoes until the Spaniards started traveling to the Americas. Most of their game was hunted to extinction. To get meat they decided to lure ET's down to earth so they could eat them. They constructed huge mounds that resemble flying saucers to entice space travelers to land. Some of those earth works like Silbery Hill and others required the equivalent work compared to today as putting men on the moon.

Mankind has been eating his gods for millennia. During communion the Catholics offer up the blood of Christ symbolized by a sip of wine and a dried wafer symbolizing the body of Christ. The word Christ means king. It doesn't necessarily have anything to do with Jesus. Jesus, however was a king of the Jews.

Why does the Pope wear a fish on his head? Ancient Greek writing talk about the Oonays who came up out of the sea and taught them many things and advanced technology.

Twelve thousand years ago a large flying craft crashed in a remote high altitude region of China known as Dropa. The local tribesmen were hunting down and killing the little occupants for food as they were starving for protein. The small occupants of the spacecraft tunneled underground to escape. There are caves in the area cut into solid rock with perfectly smooth walls as if they were cut with high power lasers. They also left behind 800 or so round discs with a hole in the center that resembled their spacecraft.

A few decades ago someone noticed that there were micro texts inscribed along the perimeter of the discs. One Russian scientist was finally able to translate some of the fine text on the Dropa stones. It resembled a form of ancient writing from India known as Erdu. It described how the astronauts came from the moon and crashed landed on earth.

Some of the local Chinese village chiefs were smart enough to capture some of the small space creature and keep them as pets. They called them the wizards. That was about the time that China invented most of the technology we have today. Gunpowder and practically everything else was invented in that area of China.

The Anunnaki or Nephilim mention in Genesis brought kingship to earth and instilled in mankind certain rules to live by. This included fifty or more commandments. The first five commandment had to do with protecting the alleged Gods from uprisings. For example, the first five of the Ten Commandments given to Moses are designed to keep the populous docile and obedient especially to the creators who cloned them.

1. <u>You shall have no other gods before Me.</u> *They seem to be self- centered.*
2. <u>You shall not make idols.</u> *They don't want anyone to make voodoo dolls.*
3. <u>You shall not take the name of the LORD your God in vain.</u> *They don't want to be talked about or made fun of or any derogatory statements behind their back.*
4. <u>Remember the Sabbath day, to keep it holy.</u> *They wanted their slaves to devote one day of the week to serve them and nobody else. This included feast prepared for them. Remember these were big guys twenty-feet tall who required a lot of food..*
5. <u>Honor your father and your mother.</u> *This could also mean their heavenly father or scientifically advanced creator god astronaut who came from space.*
6. <u>You shall not murder.</u>
7. <u>You shall not commit adultery.</u>

8. <u>You shall not steal.</u>
9. <u>You shall not bear false witness against your neighbor.</u>
10. <u>You shall not covet.</u>

> These are all good rules to live by and necessary for mankind's survival on earth but don't forget the golden rule. "Do unto others as you would have them do unto you." This woud include doing good deeds to help out your neighbors!

To protect your tribe or small nation you need warriors. It says in the scriptures 'go forth and multiply'. Obviously the Biblical patriarchs took this literally. For thousands of years the kings and Pharaohs have been impregnating as many women as possible to amass armies to fight and take from their neighbors. Kingdoms grew by killing and kidnapping wives and children of neighboring tribes. This same thing is still going on today. To expedite the process they even took their first cousins as wives. This did not help the gene pool in certain parts of the world because now there are thousands of people with low IQ's. Some of the Biblical patriarchs had as many as five hundred wives and hundreds or so offspring. Osama bin Laden, for example had more than five wives and fifty children. He probably impregnated a few servant girls in his spare time.

Marduk was instructed by his father Enki later known as Thoth to build a city unlike any other. This city was to be the center of world trade for thousands of years and it was. According to Egyptian texts Marduk learned all of the knowledge of the ancient inter-stellar beings, mathematics, design, construction using anti-gravity devices, genetic manipulation and raising the dead. The German Nazi Party was intensely interested in Marduk for their technological advancement.

During the day sunlight reflecting off the desert sands bounced off the sides of the polished-limestone, covering of the pyramids making them appear like the sun god had come down from heaven. This illusion made it look like the Pharaoh had captured the sun god to light up the geographic center of the World.

Twelve thousand years ago Earth was deep in the Ice Age with large ice caps on both poles extending over parts of South Africa, the south island of New Zealand, England and the ice sheets extended as far south as what is now Kansas in the Americas. Our solar system was in apogee in its orbit around the Sirius multiple star systems at a distance of 9 light years. At the present time we are at 8 light years and heading back toward Sirius A and B at 7.5 kilometers per second. When Ice is covering most of the world it makes it difficult to find new deposits of gold. The average global temperature was below freezing at 32 degrees Fahrenheit. People had to live near the equator to grow crops to survive. Something had to be done to replenish earth's atmosphere and thaw the ice caps back.

At that time the year was 360 days—the same number of days as the number of degrees in a circle. Ancient calendars show this. People were not so stupid that they could not count the number of days in a year. The earth was not tilted in relation to the sun and the sun shown directly on the equator all year. There were summers and winters as Earth was in an elliptical orbit around the sun. The Galactic Council agreed to put a large moon into orbit around Earth to keep it volcanic active and replenish the atmosphere with more CO_2 to warm the planet and grow more plants and trees.

After it was decided to bring the Moon into orbit around Earth Enki was forbidden to tip off the humans. There were many giants and half-breeds in the land and the Anunnaki wanted to reduce Earth's population and purify the races—much like the Germans tried to do in WWII.

Although forbidden to tell the humans about the impending flood and world-wide cataclysm Enki tipped off Noah of indirectly by whispering outside his house of sticks that there would be a world Wide flood and that he should build an Ark or boat big enough to hold his entire family and enough animals to sustain them after the flood.

It is said that Noah was a big man over ten feet tall with pure white skin like an albino—possibly because he was the son of Enki and half Anunnaki. There were many giants in the land at that time and they were consuming all the resources. Something had to be done.

Noah - Wikipediaen.wikipedia.org › wiki › Noah In the traditions of Abrahamic religions, Noah is a prophet featured as the tenth and last of the ... The primary account of Noah in the Bible is in the Book of Genesis. Noah was the tenth of the pre-Flood ... mortal life as the patriarch-prophet Noah. Gabriel and Noah are regarded as the same individual under different names.

Alleged Bible Facts from the internet:

Noah is the tenth generation descendant of Adam; Noah's grandfather was Methuselah, who is the oldest man recorded in the Bible (Genesis 5:27). He lived to be 969 years old. Some have said that Methuselah died in the flood, but most likely he died right before the events had happened. In either case, we don't have enough details from Scripture to know.

Noah himself was 500 years old when he had his sons Japheth, Shem, and Ham; At the time of the flood, Noah was 600 years old (Genesis 7:6).

Upon stepping on dry land after the flood, Noah first built an altar to God (Genesis 8:20); Noah planted the first vineyard after the flood, and he is the first drunk recorded in the Bible (Genesis 9:20-21).

Contrary to what most people believe the Ark was a very well built boat more than 800-feet long. It had wells built into it to raise and lower drogue stones with several windless. It was held together with iron bolts thousands of years before the Iron Age. It had several levels like a modern cruise ship. It also had large oak trees sticking out fore and aft under water with drogue stones attached so that when it drifted into shallow water the twenty or so drogue stones would slow it down. They heavy stones below the ship also lowered the center of gravity to the point where it was impossible to capsize it. It was a marvel of advanced engineering and seaworthiness designed to take on any sea not experienced by modern ships today. Such a ship would be able to ride out waves in excess of five-hundred feet.

I know a little about such things having fished king crab 25-years and encountered 80-knot sustained winds and 100-foot waves crossing the Gulf of Alaska. Believe me it was not fun. In fact, it was the most terrifying experience of my entire life. My old 74-foot tugboat Mary M built in 1929 drew nine feet of water. We had re-calked the entire hull in Anacortes and replace 34-planks, a new forefoot and a new bow stem at Tony Loverick's shipyard in Anacortes, Washington. Afterward I traveled to Seattle and installed a powerful new D-8 Caterpillar diesel. It was the large slow speed

D-342 Cat engine weighing 9,000 pounds. Top speed was 1,200 RPM. I did all the work myself at the end of 88 Hamlin Street in Lake Union in front of Wynn Brindle's office.

When we finally finished I took the Mary M into Lake Washington to see how fast it would go. It was so quite compared o the old engine cruising along at ten knots that you could hear the wake curling off the bow. As I drove by Bill Gate's house I gave him the finger.

We traveled north to Alaska through Lummi Passage and the Straits of Georgia stopping at Ketchikan. After leaving Dixon Entrance we stopped at Yakutat to visit Byron Mallot and others. The weather forecast was thirty knots southeast. As a king crab fisherman 30 knots on the stern was nothing to worry about. I was anxious to get going so I departed early.

The weather forecast was thirty knots southeast. I was anxious to get going and thirty knots on the stern pushing us along would be a nice ride. After we were fifty mile offshore the wind picked up to over eighty knots. I fished king crab 25 years and never saw anything like that. You could not see the size of the waves when you were in the trough.

Above is a picture of my 76-foot tug/fishing boat Mary M. It was heavily ballasted with concrete.

Each wave filled the back deck up with six feet of water while the green wall of water was in front of the boat was almost vertical. Finally the boat would rock quickly sided to side and lift up as the wave passes under us. The powerful tugboat engine running at full throttle struggled to climb the back of the green wall of water. Half way up the giant waves my boat slide backwards down the wave about 18-feet or so. I understand a little about what Noah and his family experienced.

My old radar picked up Montague Island and we were within 10 miles of Hinchinbrook Entrance the wind switch around completely to North East. The wind switch knocked the big swell down but I was suddenly bucking into 15 to 20 waves. I was overconfident after having survived the last eight hours in 100-foot waves. I kept going at full throttle. Suddenly a large wall of water came out of nowhere and smashed into the pilot house breaking two pilot house windows. I ducked the broken plate glass and was suddenly drenched in cold sea water. A foot-high wave of water ran aft through the passageway into the galley where my poor wife and children were crying. As a foot of water sloshed back and forth across the galley floor I tried to reassure them that everything was okay and that we would be inside Price William Sound soon.

After the wave broke the windows in the pilot house every piece of electronics quit working. All I had to steer by was the compass that was heaving around violently. Once I backed off on the throttle the pitching and rolling slowed a great deal but it took longer to get inside the entrance.

While idling along my deckhand Jerry nailed boards and raingear over the outside of the broken windows. After another hour or so I could see Hinchinbrook Entrance through the gloom. Once inside Prince William Sound it was still quite rough. We traveled across the sound to a small island with bay marked Snug Harbor on the chart.

I believe there were several ARKs built around the world to save remnants of humanity as every civilization has legends of surviving a great flood. Using critical thinking which we are not allowed to do Earth has undergone at least five major extinctions in the past and several minor ones in between. Given the fact that there are over 30-million species of plants and animals on the planet someone obviously has been going around seeding planets between extinctions to keep life going.

The largest Roman ship was Roman 10,000-amphorae carriers, ladened with 550 tons, were the largest ships afloat. One of the biggest of all Roman monsters was Caligula's Giant Ship. Remains found during the excavation for Rome's international airport suggest a length of perhaps 104 meters (341 ft) and a beam of about 20.3 m (66 ft.).

CHAPTER NINE

ENKI AND THE MOON

The Moon always belonged to the Anunnaki. It is a billion years older than Earth and may have been used to terra-form other planets by putting them into more circular orbits around their suns and tilt them to produce more prominent summers and winters. The resultant gravity of our Moon's force on earth of 4e Newton's per second increased volcanism keeping the interior of the planet liquid to replenish the atmosphere with more CO_2 thus increasing plant growth in the oceans and on land.

Our Moon was not formed with earth. Most of the moon rocks brought back to earth by the Apollo missions were so light that they floated in water. Their specific gravity was much less than earth rocks because they were formed in a lesser gravity. Ancient writings tell of a time when there was no moon. You can download Immanuel Velikovski's little 100-page book, IN THE BEGINNING free on my web site. www.GuardDogBooks.com

Chapter three is a translation of ancient Greek writing that talk about a time when there was no moon in the sky. Other ancient civilizations around the world have legends about no moon in the sky. We researched ancient cave paintings dating back over 12,000 years and there is no moon. Everything you have been taught in school about the subject is wrong to keep humanity in the dark about our real origins.

Apparently the Anunnaki bounced the Moon off the northern hemisphere to prolong the life of the planet with increased volcanism and more prominent summers and winters. The resulting impact killed off all the camels, horses and mastodons in the northern hemisphere and quick froze most of them in frozen gravel. Some of the animals even had green food in their mouths.

Tilting Earth and putting it out further from the sun in a more circular orbit increased the number of days in a year from 360 to 365 ¼. Having the sun's rays strike further north and south as earth orbits around the sun increased the plankton blooms and doubled fish populations. Increasing plankton blooms also increased earth's atmospheric oxygen levels to around 30% at the time of Moses. Now it is down to 20% because mankind keeps on burning things in including forests.

In my book, COSMOLOGICAL ICE AGES I used an online computer Arizona/edu /impact effects to simulate the impact. My co-author and I decided to use an angle of 11 degrees so it wouldn't bounce off the atmosphere and a speed of two kilometers per second. The Moon's mass is 7.35 e20 kilograms. The Earth's mass is 5.89 e 24 kilograms.

The resulting impact force was 750 terra-mega tons. The computer said the impact would depress Earth's crust 5 kilometers and bounce back up to 1.3 kilometers. It just so happens that the Arctic Ocean has an average depth of 1.3 kilometers.

Earth is a liquid ball with a radioactive iron core. The tremendous force of the moon impacting earth was transmitted around the iron core to separate Antarctica from South America and pushing it up to an average altitude above sea level of 7,500 feet.

Lake Titicaca in South America was pushed up 10,000 feet above sea level. The cities of Pumapuka and Tiahuanaco were completely destroyed and pushed up to an altitude of 13,000 feet above sea level. You can read more about how the Anunnaki terra-formed Earth in my book, COSMOLOGIAL ICE AGES.

More than one god walked the Earth prior to the time of Moses. There were at least eight in the Garden of Eden. The Egyptians called their God, Thoth, Ra. The Greeks called their God Hermes who practiced Hermetic Law (as above so below) and Romans had Mercury who could fly around and deliver messages at high speed. He was very fast!

YHWH said to Moses: "I am that I am." "I am Thoth." It seems that every thousand years or so he changed his name. If you are a God, you only need one name. If you want to change your name, who is going to object if you are a God?

There sending someone someplace with a high frequency microwave antenna....

Note the cell phone remote control devices in their hands. They are taking pictures of something. It looks like a couple of their cylinders need work! Below is the brainwashed human and the Tower of Babble.

God's real name was kept secret from the Hebrews when they departed Egypt during the Exodus because Moses thought the Hebrews would not worship the same god that was responsible for their slavery. Instead, Moses God was a giant lizard or bird- like entity known by the unpronounceable name of YHWH. Vowels were added to their name of God so that the Hebrew children could pronounce it. His name became YAHWEA, which was later translated into Jehova.

Some of the giant Anunnaki beings are portrayed as having bird heads on Sumerian wall reliefs. Why would the artists carve bird-like beings walking the earth with wings if they didn't really look like that? My take on the matter is they really did look like giant birds with sharp curved beaks designed to eat meat—hence the reason why Moses had to feed his god sheep, goats and the occasional human sacrifice. Giant lizards with a hard pallet cannot pronounce vowels so the real name of God actually was YHWH. It was pronounced like the Klingon language in the Star Trek movie series. Hard pallets can't utter vowels.

The Anunnaki were huge terrible giants. That is a full grown lion clinging to his leather apron. After he was done flaying you with his flail he'd let the lion loose to eat your remains.

OVER A MILLION PYRAMIDS WERE BUILT AROUND THE WORLD and we don't know why they were built.

There is over a million pyramids in Central and South America alone. Obviously the Anunnaki were incredibly advanced race with a few thousand years head start in evolution. They had the ability to clone sheep and cattle and even their own DNA to create a worker race to serve them. They had developed space travel and the ability to beam livestock and humans off planet using the inter-dimensional properties of ORMUS which I will explain later.

The Anunnaki engineered and built with their human slaves built pyramids all over the earth to create a global network of power stations. They used giant saws to cut large blocks of granite weighing up to 2000 thousands of tons. Such equipment performing such precision is way beyond our capability today. We don't have saws that can cut such large blocks of red granite. We don't know anything hard enough to cut through such stone except diamonds and even diamonds don't seem capable of doing the job. How did they cut exact absolutely square corners inside of granite boxes? We have no idea.

Scientists still haven't figured out how pyramids work or why they built them. Some think they were constructed to correct the Shuman resonance of Earth—eight hertz. The pyramid buried underground 60-miles north west of Mt. McKinley in Alaska is still putting out enough electricity to power all of Canada and Alaska. If we do understand how they work the government isn't telling us because it would disrupt big oil and the energy cartel.

The apex stone of the Great Pyramid itself was cut from a pure, glass-quartz, and crystal with gold stringers. According to Dr. Patric Flannigan: "The angle of the Great Pyramid is such that it has an impedance of 377 ohms." This is the resonance of free space." He went on to say: "If you change the angel you change the impedance." I never understood this statement the first three or four times I heard him make it on my videotape. What this means is the great Pyramid is capable of resonating in a variety of frequencies and thus transmitting energy or signals through free space. It could serve as a scalar weapon of immense power as well as a communication device.

THE WHITE-POWDER-OF GOLD

When archeologists they opened the Great Pyramid in 1909 they found a very fine white powder on the floor and stuck to the walls and ceiling.

They swept it all up and threw most of it away except for samples the sent to England for further analysis. They thought it was aluminum powder but aluminum does not exist naturally on earth as a metal or powder. It has to be smelted from bauxite ore with electricity. Supposedly the ancient people didn't have electricity???

In ancient times the Great pyramid contained a box measuring approximately two feet by three feet and two foot high. This box was made of wood, covered with gold and brass. A detailed description of how to construct this box is given in the Old Testament and the Torah. Two gold cherubim or angles are mounted on top of the lid facing each other. Between the angel's outstretched wings is a brass pan or mercy seat which was used to burn offerings with electricity.

PHILOSOPHER'S STONE

An old Alexandrine text makes particular mention of the weight of the Philosophers' stone-which it calls the Stone of Paradise. It states that: "When placed in the scales, the stone can outweigh its quantity of gold: but when it is transposed to dust, even a feather will tip the scales against it." Sir Lawrence Gardener's contends in his book: Star Fire, The Gold Of The Gods "The true Grail bloodline originated with the Anunnaki gods in southern Sumeria at least 50,000 years ago and was sustained by ingestion of an alchemical substance called 'Star Fire'."

Everyone who has ever read about or seen pictures of the Great pyramids of Giza has asked himself or herself the question: 'Why were they built?' They were laid out according to the stars and aligned with true North with great accuracy.

Once a person is aware that small viewing ports (wave guides) cut into the stone hundreds of feet long and aligned with certain star constellations and once a person knows that an inter-dimensional, super-conducting, white-powder capable of amplifying brain waves several billion times over was placed inside the pyramids then one has to come to the obvious conclusion that they are used as communication devices.

They were also built to warn mankind about cataclysmic events that could destroy all life on earth. What more important reason can there be for someone to build such monumental edifices than to warn future generations and make a phone call to the other side of the Galaxy? It is an inter-stellar communication device capable of communicating with thousands of intelligent races throughout the Galaxy. The communication was instantaneous and not dependent on the speed of light because the material in the Ark is inter-dimensional. What more important reason can there be than for an inter-stellar race to phone home? "ET phone home?"

This raises the question: "How many other worlds with intelligent races with pyramids are scattered throughout the Galaxy?" The next question that comes to mind is: 'If were so smart then why aren't we building pyramids to communicate with other worlds'? Why aren't we building monuments to warn future generations about impending cataclysmic events? The truth is we are not advanced enough to build a pyramid let alone figure out how it works and be able to predict when the next comet will hit. We are definitely not organized enough to do anything about it. We haven't even reached the point where civilization was 12,500 years ago.

Maybe in another fifty years or so if we don't blow ourselves up first in the meantime we might get it together enough to do something about our future.

An old Egyptian gold mine

In 1948 archeologists discovered what they thought was an old Egyptian gold mine on the Sinai Peninsula. They found a crucible for melting gold and a large granite table with a slot cut across the middle of it. I am convinced the half-inch deep slot was to catch fine gold when high grading the ore. As one with a rudimental knowledge in gold mining, high-grading is the process of pounding gold quartz ore to break up the quarts leaving the gold behind. Some gold bearing ores have gold wire stringers thru out. Such ores are pounded with a hammer to break up the quartz to leave a ball of gold wire. The slot in the granite table was to catch the fine flakes and pieces not attached to the gold wire.

The archeologists also uncovered six rooms adjacent to the mine. One of the rooms contained four inches of a very fine white powder. After they uncovered it the desert winds blew it away. To this day they have no clue as to what it was. I have to conclude the rooms were constructed to hold the white-powder gold.

Island of Crete the home of the Gods

The same year in 1948 on the Island of Crete known as the home of the ancient Greek Gods, a platoon of United States soldiers were digging out two twenty-foot-by twenty-foot underground rooms. The rooms were approximately eight feet deep and lined with green treated timbers. They soldiers had orders to look for gold. The rooms were packed full to the top with a very fine white powder. They threw all the white powder away. One of the soldiers that were there lives in the town of Nikiski not far from where I live now. He took some rope samples back with him.

I believe what the soldiers were digging thru was gold in the ORME state. They had discovered the actual food repository of the Gods—the most valuable substance on earth and threw it all away…

James Hurtak, Anthropologist.

According to Egyptian Legend there was a great war in Heaven. All the most beautiful women were put on board the Maricaba space ship and taken to earth. (Did the Devil take women with him when he was exiled? Hell Yes!!! Wouldn't you?) According to James Hurtak, the Hebrew word Mercaba was the name of the divine space ship that brought man to earth.

Mainstream science says the pyramids were already a thousand years old when Moses was alive. Graym Hancock discovered that the small channels cut into the Great pyramid were aligned with star constellations, which represent the sky at 10,500 BC. If this is true, then the pyramids are at least 12,500 years old, which would mean that they were about 9,600 years old during Moses's exodus from Egypt.

The oldest mummy found to date is 4,500 years old so what happened in the meantime? Did the flood kill everybody off? What happened after the great flood? Was the Nile valley resettled 5,000 years later? All these scenarios are quite possibly true but we may never know for sure.

MOSES

Practically all of us know the story of Moses. The Pharaoh decreed all the first born of the Hebrew slave families were to be killed. To keep Moses alive his mother placed him in a reed basket coated with tar and set him adrift on the Nile. The Pharaoh's daughter found Moses and raised him as her own - eventually Moses became an Egyptian priest. When Moses was twenty-five he accidentally killed one of the Pharaoh's overseers because he was whipping a Hebrew slave. For his crime Moses was banished from Egypt for life.

Moses wandered out across the dessert first north and then south following an ancient caravan rout running parallel to the Gulf of Suez. His rout took him from one watering hole to another toward the Gulf of Aqaba. He waited until there was a strong wind from the Northeast and crossed over to what is now Saudi Arabia. At that time it was known as the land of Midian. Moses met a young woman in Midian who happened to be the daughter of a Midianite Priest. He fell in love, married, and lived a quiet life of a shepherd. Moses watched over his father-in-law's sheep twenty years. Every day he prayed to God for his people's salvation.

One day God spoke to him from a burning bush. The bush glowed like fire but was not consumed by flames. Moses had become a true priest, closer to God and with all the shamanistic powers. God told him that the time was right to lead his people out of bondage: "Go back to Egypt for all the men who were seeking your life are dead." Moses had not forgotten the secret teaching of the Egyptian priests and was closer to Thoth who helped him to command the forces of nature. In his mid-forties Moses felt that his whole life had basically been a failure. Then he realized that there is a time for every event and all future events are foretold. There is an exact time to plant, a time to start a business, and a time to sell a house. If one knew exactly when to act then one was successful.

God said to Moses: "Go and lead your people out of bondage." With the help of YHWH Moses brought down ten plagues to convince the Pharaoh to let the Hebrew people go. One plague rained toads, then all the water turned to blood, and then it was locusts. The Pharaoh still would not let his one million Hebrew slaves go. If you tally up all the people mentioned in the Book of Numbers it comes close to one million people. After the first-born son of every Egyptian family was killed the Pharaoh finally let the Hebrew slaves go out onto the desert for three days to worship their God.

The ten plagues had decimated the Egyptian population so the Egyptians were slow to respond. After the three days they became suspicious. By the time they were able to get an army of twenty thousand foot soldiers together Moses and his people had a sixteen-day, head start. The million people were running for their lives. The bible mentions specific landmarks where they stopped to get water. Between the Saudi coast near the strait of Tiran and the Town of Al Bad (the biblical Elim) they stopped at the spring of Murah. (Exodus 15:22-23).

THE ARK OF THE COVENANT - THE MAJIC BOX

"An army that caries the Ark of the Covenant before it is invincible."

It must have seemed like magic to the ancient Hebrews when their priests would generate an electric ark (ball lightening or shikanna glory) between the outstretched wings of the golden cherub on lid of the Ark. To do it by using thought power was even more miraculous and mystifying.

There are several reports of the Ark being carried into battle. It was customary to have four people carry the Ark on long poles slipped through the brass rings attached to the side of the gold and wood box. When out in the open one could not touch it for fear of being electrocuted because it gathered energy from the atmosphere. The MANNA or ORME material inside also had thousands of amps of electricity flowing through it.

There is no way four men could carry the Ark over rough terrain without breaking the thin poles unless it was an anti-gravity device. Estimates of the weight of the gold and brass lid place it at 500 pounds. The total weight of the Ark itself would have been six to eight hundred pounds depending on what was placed inside yet it is described as floating along, sometimes carrying the people who were transporting it. It was customary to carry the Ark at least 800 yards or more in front of the Hebrew army. Once within range of the enemy a priest would concentrate on the manna inside the Ark and generate the 'shikanna glory', commanding the opposing army to lay down their arms and go home. The commands echoed magically inside the heads of the enemy as the Ark advanced. If you can produce an electric discharge by using thought power what you are doing is amplifying your thoughts a billion times. This must have been quite terrifying for any opposing force. Those who resisted which were few; died of a sudden heart attack as their bodily functions were arrested by the powerful force emanating from the Ark.

The Ark was a powerful instrument of war yet it could be used for peace. Gold dust or nuggets placed on the brass pan or mercy seat between the outstretched wings of the cherub was transmuted into the manna at 5,000-degree temperatures. This was the powerful manna, the food of the gods.

The Ark of the Covenant, mentioned in the Bible throughout the Old Testament, symbolizes the agreement of faith made between God and the Israelites at Mount Sinai. The Ark supposedly contained the Decalogue (the Ten Commandments given by God to Moses), Aaron's rod, and a pot of manna. The Ark represented God's presence to the Israelites, who carried it with them while wandering through the desert and into battle. The Ark had disappeared by the time of the fall of Jerusalem in 586 BC.

It is believed Egyptian priests would put a small amount of this same powder in the priest bread or manna they ate each day. The Priest bread was a course unleavened bread that takes a long time to digest thus taking it through the stomach where it can be absorbed by the lower intestine.

The white-powder is a superconductor made of gold and platinum atoms known as ORME material, (orbitally rearranged monatomic elements).

G. Pat Flanagan, Ph.D.

Dr. Flanagan describes the Ark in his book Pyramid Power: "The ark of the Covenant is described in Exodus 25:

10-21 as being a large box made of acacia wood and lined inside and out with a layer of pure gold. The description of the Ark is of a very large electric capacitor. If an Ark as described in the Bible was built and placed in the earth's magnetic field and charged it would have a charge of some 600 to 1000 volts and would have enough stored charge to kill anything that touched it. The Ark is supposed to have had the power to kill, by some mysterious energy. I believe it killed by electric charge. The size of the Ark is exactly the same as the stone box coffer in the King's chamber of the Pyramid of Giza."

What Dr. Flanagan didn't know was that the Ark contained a white powder with thousands of amps of electricity flowing through it that could be manipulated with thought power alone.

Dr. Flanagan also states that the electric gradient on the earth's surface is 200 to 300 volts per meter. According to NASA's long wire tether satellite the total atmospheric charge between the earth and the ionosphere 60 miles above the earth is between 3,000,000 to 12,000,000 volts and a current of 1,800 amps. There are 300 thunderstorms going on at any one time all over the earth. Experiments have shown that this electric charge is an important beneficial factor for all living things on the earth.

SIR WILLIAM SIMENS

The British scientist, Sir William Simens first discovered this mysterious energy signal in the 1800's. While touring the King's chamber Mr. Simens took a wine bottle and wrapped it with a wet newspaper turning it into an electric capacitor. The bottle gave off Sparks as it collected energy.

An excellent book, "THE GOLD OF EXODUS" by Howard Blum is about this subject. This book is available from: Adventure Unlimited Press PO Box 74 Kempton, Illinois 60946

The book is about Jerry William's and Bob Cornuk's modern-day, Journey to find the true Mt. Sini, which is Jabal Al laws, located in the land of Midian, which is now Saudi Arabia. They discovered the underwater bridge where Moses and his people crossed the Gulf of Aqaba to get to the land of Midian. They discovered the springs of Elam where there are twelve springs and seventy palms. They took photographs of the caves of Midian where Moses lived with his wife and children. They took pictures of the mountain of God and it was burned black at the summit as if some holy fire had engulfed it at one time. They also photographed the altar of the golden calf and drawings of cows on the rocks. There are no cattle in Saudi Arabia unless the fleeing Hebrews brought them.

Before Moses departed Egypt he enlisted his older brother Aaron and two friends to go inside the Great Pyramid via the secret door and take the Ark. The ten-ton, door was hinged in such a way that two men pushing in the top could open it. Moses took the Ark because it had been used to enslave his people. Inside the Ark was the Egyptian Book of The Dead in which God already wrote the Ten Commandments. Also carried in the Ark was the quartz capstone of the Great Pyramid.

Moses wanted insurance that his people would never be slaves again so he took all the sacred objects and several tons of gold as well. He actually ripped them off as pay back for their captivity. From that moment on the old Egyptian rituals were forgotten. What we didn't know until now is that the Ark also contained the white-powder-of-gold or manna.

The million or so Hebrews didn't stop to worship their God. They kept going east then South along the coast of the Gulf of Suez then east toward the Gulf of Akaba. They were fleeing for their lives so they didn't waste any time. Moses knew where there was a secret underwater crossing used thousands of years by peoples long before the Egyptians. Moses had studied all the Egyptian texts and knew where he was going besides he had been their once before. He was heading for the fabled mountain of God where he used to graze his uncle's sheep during his long Exile from Egypt.

GENETICS

Genetic manipulation of the human race is evident in the story of Moses— leading his, 'Chosen people' out of Egypt to the Promised Land. Moses took the Ark and fed his people manna or priest bread made from a small amount of the ORME gold to keep them alive. Nobody, not even whole armies could follow them into the dessert because they couldn't carry enough food and water.

How did the Hebrew do it? It seems that people who fast then ingest a small amount of the most nourishing substance on earth, the white-powder-of-gold baked in bread do not need much food to keep going. It was either keep going or be killed by Egyptian soldiers. We're not talking colloidal minerals here. Were talking about a priori state of matter that is easily transmuted into other elements by our bodies, as it is needed. For water they licked the due off from the leaves of plants and dug up roots. God, through Moses provided manna on several occasions when they were in dire need. Moses caused the Hebrews to wander in the wilderness forty years to kill off the old people thus wiping out the many old religions. He was pushing monotheism at the time.

The Hebrew children who grew up eating the Priest Bread had a different, genetic makeup than the old generation and most importantly, they believed in one God. Their belief system is what held them together as a race of people for thousands of years. This is the reason the Hebrews could live in other nations with other cultures and religions for centuries and still remain Hebrews. After all this time, they still believe in the one God. They couldn't have done it without Moses.

According to the Bible Moses prolonged his own life span living to the ripe old age of 120 years. This is a remarkable accomplishment considering that the average life span at the time was thirty years.

The band of Hebrews traveled south to the narrow entrance of the Red Sea where the water is approximately ten feet deep. Moses and the Hebrew travelers knew there was a passage at low tide. Napoleon traveled this same passage many years later. Napoleon's engineers discovered it in 1799. A strong northeast wind causes the passage to go dry. Napoleon barely escaped with his life when the wind stopped suddenly. As soon as one can see dry land you run to the other side as fast as you can before the wind stops blowing. The Red Sea passage is slightly over five miles wide at this point. The ancients had been making this trek for thousands of years and Moses probably knew it existed from temple writings or possibly word of mouth. He had traveled the same rout before. The ancient caravan trail leading up to the edge of the Red Sea on both sides of the passage is clearly visible in infrared, satellite-photographs today.

After Moses came down from Mt. Sini the Bible tells us he ordered his alchemist and goldsmith Bazaliel to burn the golden idols into the white-powder. The last vestiges of the pagan religions were consumed by fire. Moses then threw this white powder into the stream and made his people drink of the waters thereof. This was done to enlighten because they had sinned by worshiping the golden idol. The records of the mystery schools cite rather more precisely that shewbread was made with the white-powder of gold. While Moses was up on Mt. Sini the Hebrew God YHWH (Thoth) dictated the first five books of the Bible known as the Torah. For forty days and nights Moses copied the 304,805 Hebrew letters in the books of Genesis, Exodus, Leviticus, Numbers, and Deuteronomy. Thus YHWH or Thoth gave mankind it's first written history in a form of codes written within codes. In these first five books of the Bible the coming of the savior Jesus Christ is predicted and discussed in detail.

ARCHEOLOGISTS, SCIENTISTS, AND ORDINARY PEOPLE

HAVE DISCOVERED AND LOST THE SECRET OF THE WHITE POWDER-OF-GOLD HUNDREDS OF TIMES. In 1948 Archeologist uncovered an Egyptian temple at the base of Mt. Nebru. This name is awfully similar to Nebiru the planet where Marduk is suposed to have originated. This is not the same mountain as Jabal Al Laws. The ruins of the ancient Egyptian temple are near the base of a Mt. Nebru. Inside the temple were many split-level, tables of different heights for high-grading gold ore. High grading is simply a process of pounding the quartz ore with a stone hammer until it is all broken up. The gold and gold stringers are picked out of the crushed rock and collected. If the sample is very rich the gold stringers actually hold the gold together forming a lump of wire.

The temple had six storage rooms off to one side. Inside one of these storage rooms was four inches of a pure, white powder. The Archeologists took samples of this white powder and had it analyzed but again the tests were inconclusive. After the temple was uncovered that dessert winds claimed the rest of the white powder. When it dries out it becomes very light and floats away. I mention this because it shows that the Egyptians had been mining gold in the Sini for millennia.

DAVID HUDSON

David Hudson, discovered (I should say re-discovered) the manna from heaven in the 1970's. After seven years of extensive research during which time he spent over one million dollars of his own money. David Hudson began making fuel cell panels out of the platinum-group-metal, rhodium extracted from his mine. He sold them to General Electric to be used in the fuel cells, which furnished electric power for the space shuttle and other applications.

General Electric, said to Mr. Hudson: "We are paying you for Rhodium but the panels don't test rhodium. They work just fine in the fuel cells but what are they made of?" David said: "They are rhodium but in a different state of matter." After they used them in the fuel cells for a time and took them out they did test high in rhodium.

General Electric said to David: "The first one to understand this should get the patent on it." GE subsequently helped David secure 22 patents on Orbitally Rearranged Monatomic Elements. David Hudson spoke in his February 1995 question and answer session: "...First of all it is a room temperature superconductor! When mixed with water it forms a gelatinous mixture. When ingested, it had the following affects. Every cell in your body will be taken back to the state it is supposed to be, when you were a teenager or a child. It perfects the DNA, and closes the light within the body until you literally reach a point where the light body exceeds the physical body." "The gifts that go with this are perfect telepathy, you can know good and evil when it is in the room with you, you can project your thoughts into someone else's mind, you can levitate, you can walk on water, because it is flowing so much light in you, you literally don't attract to gravity."

When gold and platinum is in the monatomic or double-atom, state the electrons of this material are Cooper paired passing each other double the speed of light to balance the atomic forces within the huge nucleus. Additionally electrons don't make complete orbits around the nucleus. They go ¾ the way around and disappear into hyperspace. You see electrons take on energy each orbit therefore can go on forever and never wear out. In other words everything around you including yourself is dipping in and out of hyperspace picking up energy to go on forever. This gives the material inter-dimensional properties, especially when the electrons exceed light speed.

In other words the material can exists in one or more dimensions at the same time depending on temperature and electrical excitation. This exotic matter is a superconductor carrying thousands of amps of electricity but no voltage unless excited by something or someone with a similar magnetic field. It has a magnetic or Meissner field of high frequency that can be energized by thought. Two of David Hudson's patents can be found on the Internet: David Hudson's British Patent is located at: http://www.monatomic.earth.com/david-hudson/patent-us-l.html

David Hudson's Australian Patent is at: http://www.monatomic.earth.com/david-hudson/patent-oz.html.

During the early 1990's David Hudson toured the United States giving lectures and workshops about the white-powder-of-gold. Portions of three of David Hudson's lectures are available on the Web. The most complete of these transcripts is his Dallas lecture and workshop. You can find this transcript at: http://www.monatomic.earth.com/david-hudson/1995-02-dallas-toc.html

Three other Hudson transcripts can be found at:
http://www.cris.com/-Notnorml/alchemy2.htlm
http://www.eskimo.com/-billb/freenrg/hudson.txt
http://elaine.telport.com/-boydroid/gold.html

David Hudson's interview of February 1995: "Okay, so in the Ark of the Covenant was the pot of Manna, and the stone through which god spoke to Moses. Now when Moses was up on Mt. Sinai, he was smelting the Manna to the gold glass. And why do I say that? Well, when you understand that in old kingdom Egypt these were called bushes not eyelashes. These were bushes: check your Egyptian literature. The burning bush was the enlightened Third Eye. And so Moses on Mt. Sinai, when he looked upon the burning bush, was a mistranslation, it was the third eye was open and God communicated to him through this stone. God didn't write on the stone, if he wrote on the stone why would he put it in a box, seal it up and let no one look at it? If God wrote on it he would put it up on a wall where everybody could read it. And when you realize that in Old Kingdom Egypt, that on the holiest day when they carried around the "sign of the Seal" you read that they carried around an Ark on two poles, and in the Ark was a stone. Coincidentally what was in the Ark of the Covenant? A stone, the Manna, the gold glass, staff of RA, and possibly the Rod of Jacob were in the Ark.

And I'm saying that round the Ark of the Covenant was the Meissner field. Now, the strange thing about a Meissner field is other Meissner fields that oscillate at the same frequency, can enter that field and not perturb it. And so if you are a high priest, a Melchizedek priest, and you eat this

Bread of the Presence of God every week, you are a light being, and you can enter into that field and approach the Ark of he Covenant and not perturb it because you are in resonance with it. But if you are an ordinary soldier or a person who thinks bad things, ha-ha-ha, you know, they have to tie a rope around your legs because as you approach it, it may have a flux collapse. Now if you can imagine several hundred thousand amps and now you have volts, it's like a bolt of lightning. It literally is energy that is of unbelievable magnitude."

EVIDENCE OF ANCIENT CIVILIZATIONS

In England there are ancient lay lines built by the Druids that travel for hundreds of miles in an exactly strait lines. Many of them were used for roads. We do not know what they are used for today. Possibly they are some kind of earth energy grid. According to Cambridge scholar Collin Andrews the lay lines completely covered much of the Earth. Mathematician, Carl Munck discovered all of the ancient monuments including the ones in South America and Australia are tied together using degrees, radians and functions of degrees of Longitude. All of the architects used the 360-degree circle and all used 60 minutes in a degree. All ancient monuments all over the world were constructed exactly on predetermined functions of radians using the great Pyramid at Giza as zero degrees Longitude. In Grand Mitneer Breeze, England the Druids erected a rock stele 150 feet high weighing 340 tons. So-called primitive men accomplished this. There is no crane on Earth capable of erecting a stone that large today. Under the sea there are raised roadbeds running for thousands of miles from the Bahamas to Newfoundland and West toward Cuba. These roads built at a time the ocean was five hundred to a thousand feel lower. They are difficult to detect by divers or submarines because they are a mile wide. Only recently have satellite radar images detected them. Obviously the earth was home to several advanced civilizations in the past.

MOSES NEVER DIED

As far as we know Moses never died. He simply walked up a mountain into the mist and disappeared. Moses was a great seer and magician. He was also a mystic who understood and acted upon cosmic and universal laws. Trained as a Hyksos Pries he was able to turn his staff into a snake, and performed many other amazing feats because he knew certain cosmic laws that were handed down to him by the Pharos.

Jewish historians say that ambitious politicians murdered Moses, but it may be that Moses simply faded away when the time came for him to vanish. In the book, Antiquities of the Jews the author Josephus writes: "As he went to the place where he was to vanish from their site they all followed after him weeping, but Moses beckoned with his hand to all who were remote from him, and then to stay behind in quiet. All those who accompanied him were the senate, and Eleazar, the High Priest, and Joshua. As soon as they were to come to the mountain called Abarim, Moses dismissed the Senate and a cloud stood over him on the sudden, and he disappeared."

Some theologians speculate that Moses ascended into Heaven, just as Christ did. However, there is some debate regarding that issue. H. Spencer Lewis, founder and first imperator of the Rosicrucian Order, believed that the cloud referred to be a mystical cloud and that Moses and Jesus actually disappeared, possibly into another dimension. There is no doubt that we all owe Moses a great debt of gratitude. Without Moses we would not have the Bible, the Torah, or the Koran. Without Moses we would not have the basic history of our creation and the Ten Commandments. Without Moses there would be no Moslems, Jews or Christians. Moses actually liberated the world from a decaying mind-control society. Now we have a consumer-based society and the Democrats are trying to go back to the failed Communist system of government responsible for killing off more than 300-million people .

CHAPTER TEN

ET'S AND NON-LOCAL VISITATIONS

NON-LOCAL VISITATIONS

In all the old religious literature there are beings, not human that are interacting with people. They are the spirits and the descendants of the Anunnaki Gods that created us. We walk among our brothers and sisters who are fellow spirits. Spirits are non-local and can travel at will to anyplace on earth to observe and help one another. Out of body experiences are real.

Gabriel is one of the EL or light beings mentioned in the Bible.

EXTRATERRESTRIAL VISITATIONS

Practically every culture on Earth has stories and legends about extraterrestrial beings coming down to Earth and interacting with humans. The Balinese say that seven celestial nymphs came down to earth from the Pleiades. A Balinese prince became enchanted with one of them and took away her sarong so that she couldn't return to the stars. He refused to give back her sarong until she had bared him a child. Today that child is all children. When Balinese children become terrified of the dark they are taken outside and showed the stars and told: "That is where you came from."

ARE BOUND HEADS EMULATING ALIENS?

Many ancient cultures bound the heads of their young. The Flat Head Indians of the Dakotas and the Indians of Eastern Washington bound the heads of their young to a board. Archeologists and anthropologists haven't a clue as to why this was done. Recently I discovered that some

Egyptian royalty bound the heads of their young to produce the high sloping forehead. Nefertiti the most beautiful and famous Queen of Egypt and King Akhenaten had the elongated skulls. Recently I was studying life-like, paintings of their profiles on stone in a book that I have on Egypt and I couldn't help noticing how much they look like the gray aliens described by so many flying saucer alien encounters. Their eyes were painted large and slanted back as were their elegant royal heads. The idea suddenly flashed into my head that these ancient people were emulating the Anunnaki gods themselves.

LLOYD PIE

Lloyd Pie is bucking the scientific establishment by arguing for extraterrestrial altering of the human specie 330,000 years ago. A new fossil was found of Homo Robustus. It has the pre-human characteristic mouth. The teeth and jaw structure clearly link it to modern man. It ate fruits, vegetables and occasionally meat. There are two marks on the one of the femurs that shows it was eaten by a large cat. The thick right humorous bone show that it had tree-climbing ability.

Homo Robustus are five to ten times stronger than us. We know this because their bones are ten times stronger than ours. Their muscle mass would have been comparable to the size of their bones. Their brain was much smaller and their foreheads were flat.

This particular find is eight million years old, which leads us to believe that the specie itself would have to be at least ten million years old because it takes several million years to develop from whatever it was before. The specie remained much as it was for millions of years, then boom! Modern man appeared 350,000 years ago. Cro-Magnon man is a completely different human being with some fossil record. The establishment is putting off the day of reckoning when it will have to revise its history books. Where did we really come from? We didn't evolve here! The truth will win out eventually.

WHY GOD CREATED MAN

ZECHARIAH SITCHIN

This famous author who translated the Ancient Sumarian texts says the pyramids were built as beacons for space travelers. The seventy-ton, granite, spirit-stones above the King's Chamber are still emitting a strange electric signal. The pyramid is sending modulated gravity waves strait up in a cone configuration above the apex. As the earth rotates this signal is beamed out in a circle every 24 hours. Far out in the cosmos space travelers would receive one brief signal every twenty-four-hours identifying it as earth.

According to Zecharia Sitchin's translation of ancient Sumarian texts the Anunnaki Gods came down to Earth from Heaven to mine gold 450,000 years ago. They tried to extract it from sea-water in the Pacific Ocean but it didn't work well and the supply wasn't enough for their needs." The gold was needed for the genetic manipulation of the many species of both plant and animal life on Earth and not just for their own use to eat. Ice covered most of Europe at that time and the climate was much colder than it is today. Earth was in the Ice Age and our solar system was at apogee in its orbit around the Sirius multiple star systems. Sirius A and B aerated most all the coal, oil and limestone on earth.

The sun could not possibly get through ancient earths 750-pound per atmosphere inch atmosphere which was 2,800-miles deep. It was all taken down with photosynthesis by light that could not have come from the sun. (*Hint, Sirius B, the size of earth puts out more than 100 times the light of our sun in the UV spectrum of 350 to 400 nanometers.) Now our atmosphere is only 14.5 pounds per square inch at sea level and fifty miles deep so we are in much more danger of incoming meteorites wiping out all life on earth. The atmosphere was taken down by photosynthesis with light and it did not come from the sun…

Ice ages currently last105000 years. After our solar system was blasted out of its orbit following the little Sirius B around Sirius A 2.5-million years ago it started making elliptical orbits. As we go further and further out away from the Sirius System the Ice Ages increased from a few hundred years to 105000 years where most of the earth gets covered with ice.

This elliptical orbit cycle shows up in the Antarctic ice core graphs where the peaks of the graphs get closer together as you go back in time. This information is all in my large astronomy book, COSMOLOGICAL ICE AGES.

After the southern hemisphere became more ice bound where they couldn't mine gold in south Africa, the Anunnaki gods traveled further north to Mesopotamia and to set up camp between the mouths of two large rivers known in later times as the Tigris and Euphrates. This camp they called Eden, which modern Biblical scholars translate, as mountain. Mainstream science more or less agrees that Mesopotamia was the site of ancient Eden.

Trying to understand the inter-dimensional material of the Gods may seem complex but this ancient knowledge is older than the human race. This exotic matter is the gold of the gods, the manna. About 35,000 years ago modern humans were cloned by the Gods using some of God's own DNA.

The Gods took their slaves and settled on the mountain of Eden because there was gold there. After all, gold is what they came for. Lacking a mechanical means to extract the gold, they tried using some of the indigenous population as laborers with unsatisfactory results. The Anunnaki use a form of telepathy or mind control to communicate with their subjects but when you don't have much of a mind to understand what needs to be done it doesn't work very well. We have the same problem today with some of our employees today. It seems that all they want to do is suck up a 12-pack and watch football all day. Given the mass produced food we eat today, is it any wonder why we only use on average eight percent of our brain capacity?

Finding the area populated by aborigines that were too stupid to use as laborers they decided to create a new type man, a high-bread using some of their own DNA. This was the obvious solution to the problem. Create a new type of man, one who is small enough to tunnel underground and smart enough to take direction, yet still retain some of the strength of Homo Robustus. Using several female eggs from various local indigenous females they injected their own DNA into the eggs and placed them into a nutrient solution, which contained a small amount of the white-powder-of-gold.

The eggs started to grow doubling the number of cells every minute. First it was two cells then four then eight.

The incubation was done to ensure that the cells would replicate properly without any genetic problems. After the eggs reached a certain size they transplanted them back into the womb of the female aborigine. The result was the first perfect man, Adam. "And the Lord God formed man of the dust of the ground and breathed into his nostrils the breath of life; and man became a living soul." -Genesis. The dust of the ground was actually translated wrong from ancient Hebrew. The correct translation is: "That which is of the Earth or Earthling!"

Genetically compatible females were created much the same way using some of Adam's own DNA taken from his rib. Each time the manna solution was used along with prayer to ensure that the cells would replicate perfectly and in the proper order. When Adam was one month old they took a portion of his rib bone to clone a compatible human female for him. The first woman, Eve was born. "She shall be called woman because she was taken out of man."-Bible. Genetics is the most awesome power in the universe. At the present time the Chinese are clowning cattle and sheep.

Eighteen years later the serpent being spoke to Eve and gave her some of the white-powder-of-gold to eat thereby raising her conscience awareness and intelligence. In other words she grew up. She objected to what was being done to humans and told Adam.

> Genesis 3, 6, "And when the woman saw that the tree (of knowledge) was good for food, (brain food) and that it was pleasant to the eyes, and a tree to be desired to make one wise, she took of the fruit thereof, and did eat, and gave also unto her husband with her and he did eat." The two of them were expelled from the mine site known as Eden because they knew too much. I think that since they became enlightened they became aware that they were being held captive to be used as slaves later on so they decided to escape.
>
> Genesis 7: "And the eyes of them both were opened, and they knew they were naked; and they sewed fig leaves together, and

made themselves aprons." Obviously Adam and Eve were raised to a higher conscious level with the manna to where they became aware that they were naked. Their intelligence was raised up to the point that they were more like their captor Gods."

Genesis 4, 22: "And the Lord God said, Behold, the man is become as one of us, to know good from evil: and now, lest he put forth his hand, and take also of the tree of life, and eat, and live forever." They may have discovered the power of the manna to increase intelligence.

Genesis 4, 23: "Therefore the Lord God sent him forth from the Garden of Eden." Enki may have let them escape from Eden on purpose because he knew his fellow space travelers were becoming increasingly evil so he sympathized with their plight.

Adam was taken from the earth and not made of earth. In other words he was an earthling. Both Adam and Eve were denied the knowledge by which they would live forever, however they were conscious of their existence. The damage was done.

Genesis 5, 5: "And all the days that Adam lived were nine hundred and thirty years and he died." He obviously had the technology to live a long life.

Genesis 6, 1,2,3, "And it came to pass when men began to multiply on the face of the earth, and daughters were born unto them, that the sons of God saw the daughters of men that they were fair; and they took them wives of all which they chose. And the Lord said, "My spirit shall not always strive with man, for that he also is flesh: yet his days shall be a hundred and twenty years." The secret to living a long healthy life was taken from them.

I interpret this to mean that there were many sons of God or Angels. The sons of God could be half-breeds or it could mean that the Angels were actually other 'sons of God' that were not born on Earth. The sons of the Gods traveled to all parts of the world, settled down to create all the various human races.

These were the giants and heroes of old. God limited mankind's life span to 120 years so that they wouldn't over-populate the earth. Let us assume for a moment that someone with god-like powers who has the ability to exit his body and astral-travel at will to any location and interact with the physical then it is easy to see how one could manipulate genes during cell division. Intent and emotion affect the outcome of any experiment.

ORMUS WHITE-POWDER-OF-GOLD FOUND IN MUMMYS

In the seventeen hundreds powdered mummy was used as a medicine. Mumia was found on the shelves of apocatary stores in Europe for many years. Doctors prescribed the mumia when all else failed and quite frequently the patient was cured. The monatomic minerals in the mummy bones did the trick. The excess white-powder-of-gold in the scull is the reason why it was prized above all else for its medicinal and conscience elevating value.

In the book of Genesis there are over a hundred people mentioned who lived over nine hundred years. The Anunnaki Gods Enili and Enki took the white powder of gold to keep their bodies functioning perfectly thousands of years. Living that long is incomprehensible to humans.

According to Peter Moon's book, Black Sun, the Nazis were planning to keep Hitler alive nine hundred years so that he could rule the world longer than the Roman Empire. A team of fifty Tibetan monks administered a substance called occultum to Hitler. This black subsistence oozes out of the ground near the presence of gold deposits. It is composed of monatomic gold rhodium and platinum that is being transmuted into the metal by Mother Nature herself. Many thousands of years ago the Tibetans noticed that dogs and monkeys licked black ooze from certain rocks. The Tibetans have been using occultum for thousands of years. It enables them to transcend death and access higher planes of existence.

The Hizenberg Uncertainty Principle states: " That whatever you study you change." There is no such thing as a chemical reaction without some kind of interaction by the person performing it, if it be procreation or a laboratory experiment. Part of ancient alchemy was prayer or meditation.

One had to hold his or her mouth right to achieve the desired reaction. The same goes for the fertilization of a human egg. The process evokes an emotional state of mind, which affects the outcome of the conception thus the reason being that the White-powder-of-gold responds to the frequency of thought.

Many thousands of years later after Atlantis, the Egyptians hired the Hebrews to change gold into the monatomic state using secret methods passed down by the gods. The word alchemy hints of the way by which it was done. Alkaline was skimmed from off the top of the soil in the Nile Valley. They put the alkaline into a pot with a little gold ore and boiled it several days. A small amount of the ORME gold was in the Egyptian soil was dissolved in by the Alkaline. Plants use the same process when they grow. Plant extrude alkaline from their roots to dissolve the minerals and other nutrients which they take up into their leaves to convert into energy food using photosynthesis.

The toughest metallic substance known to mankind on earth is rhodium. Mankind cannot go into space without rhodium and the other six platinum group metals because it is used in rocket engines and wherever extreme strength and heat resistance are necessary. Rhodium can be detected with a spectrographic analysis machine capable of a 400 second burn. The long burn time is necessary because you have to first boil away all the other elements. Everything has a boiling point and a flash point.

American spectrographs have three-eights'-inch diameter carbon electrodes get to the iron then the carbon electrodes are gone and the sample falls on the floor. It takes one-inch, diameter carbon electrodes and 900 amps of electricity to even get to the point where the platinum group metals start to flash into nothingness. Rhodium is the toughest element on earth and can survival 386 seconds in an electric arc at which point it will flash into nothingness.

[I should explain to the reader that spectrographs photograph the light given off by the electric ark using a revolving film strip which is exposed through a slit and a filter. Whenever a given element reaches the flash point it gives off a burst of light which is more than the carbon ark itself is giving off.

This flash leaves exposed strips on the film. Where the flashes are located on the film correspond to the various burn times thus enabling the operator to determine which kind of elements in the sample.

A large flash signifies a larger amount of any particular element in the sample.

By adding up the weights of all the elements approximated by the test one can estimate the percentage of ORME material in the sample. All the elements are accounted for; Right? What else is there? What could possibly be left over after all the elements are gone?

Due to their high-speed electrons ORME atoms are very light and can float away. Even a spectrograph won't detect them. This is the fabled anti-gravity material that mankind has been searching for. A small amount of ORME atoms help the alchemy process along turning the rest of the gold in the pot into the ORME state.

It is alchemy where the high frequency ORME atoms raise the frequency of the other material. This is the fabled philosopher's stone. After boiling the Egyptian soil in alkaline solution for a week they poured it into a vat where the monatomic atoms or ORME material was precipitated out by adding a little wine or vinegar to raise the acidity. The powder then became visible in the solution and was filtered out and dried. A hundred Hebrew slaves performed this task daily to keep one ARK full. Several Arks were kept inside the various Pyramids around the world and used as high-frequency communication devices and possibly to transport human priests from one temple to another.

The pyramids were never intended to be burial tombs. No bodies were ever found inside the large pyramids. The pyramids were tremendously powerful terrestrial and intergalactic communication devices and they were used to control men's minds. I admit that some hocus-pocus probably took place with the use of various static-electricity devices to frighten the simple folk into compliance. Included were rods and staffs that could take a charge and shock people. A large globe sitting on top of a Van de Graff machine was used to light interior passageways. The Egyptians knew about electricity, as the Nile is full of a kind of electric fish known as Gymnarchos Niloticus.

According to mainstream scientists and archaeologists, the Egyptians compensated their workers with bread and beer. I don't concur with this assumption. Surely they had other forms of compensation especially for priests and high-ranking bureaucrats. I am willing to bet the engineers who oversaw the construction of the pyramids were highly compensated. A promotion or higher status in society would be considered compensation for a job well done. A promise of eternal life held even greater value. Keeping a work force of such size as to build the Great pyramid would have been impossible using bread and beer. Unlike other scholars I think spirituality had very little to do with keeping them working. I don't think they were any more spiritual or ambitious than the average worker of today. My contention is that they were kept working as slaves with a mind-control device of extreme power known as the Ark.

CHAPTER ELEVEN

TODAY'S MIND CONTROLL

Eating small amounts of the gold of the gods in bread can extend the life spans of individuals and increase intelligence. This information has been the world's best-kept secret for thousands of years. It also increases the amount of electrical activity in the brain, allowing neurons to super-conduct electricity at tremendous speeds. A person goes from using ten percent of their brain to using eighty or even ninety percent within a month of ingesting the white-powder-of-gold.

The psychological effect of such a transformation is not for everybody. Suddenly a person becomes psychic, reading other people's minds and knowing all truth. Some people cannot handle the information input and go off the deep end so to speak. Others may become extremely self-riotous and superior in attitude, thinking that they are gods that know everything. Obviously this isn't the case either. The transformation can create psychotic and paranoia episodes.

For some unknown reason the information on this subject started to come my way. I live in Tuxedni Bay, Alaska between two active volcanoes. Tuxedni Bay is located on the west side of Cook Inlet approximately fifty miles southwest of the city of Kenai. My parents came here in 1942 to process salmon. My father had constructed a floating cannery on a power barge. They fished and canned salmon for a living.

The other reason I live here is because when I look up at the stars at night I don't see any industrial haze. I like the independence, the freedom, the clean air, pure water and natural foods that abound here. I learned that it really doesn't matter where a person lives. No matter where you go there are people living and prospering. It all depends on how hard you are willing to work.

I notice more and more people traveling all over the world searching for a better life and most likely never finding it because they are looking for the easy way. They have what is known as the Santa Clause syndrome. To be happy all they need to do is grow up and go to work. Do what you want to do but get on with it. Over thirty years elapsed before I started to discover the true meaning of life, the white-powder-of-gold and how it relates to the history of man and how it could be used to help mankind.

David Hudson spent over a million dollars of his own money doing thousands of spectrographic analysis tests in his Phoenix, Arizona laboratory. Mostly he was interested in soil. Only later did he test the leach solutions from his gold mine. Eventually he discovered that he had been throwing away thousands of dollars of rhodium and platinum.

When he was working for General Electric they told him that the first one to understand this material and describe it scientifically should get the patents on it. David Hudson filed twenty-two patents on orbitally rearranged monatomic elements.

Mainstream science still refuses to admit that the ancients had more knowledge about the physical universe than we do today. The truth is, modern science was totally ignorant of the white-powder-of-gold found inside the great Pyramid and the Ark and they have no idea what it was used for.

When archeologists opened the Great Pyramid for the first time in 1909 they discovered a great deal of this powder inside the King's chamber. It had floated out of the stone box where the Ark was kept and it coated the walls and ceiling. The archeologists swept up the white powder and sent a sample to London England for more tests. The tests came back as silica and aluminum powder---this at a time when aluminum was supposed to be unknown to the ancients. After conducting a few tests on the white powder they threw it out. All the samples were disposed of because the tests were inconclusive. They didn't have spectrograph that could burn long enough to detect the platinum group metals let alone figure out what the ORME state was. They threw away the most valuable substance on Earth because they thought it was radioactive sand. So much for modern science!

David Hudson spoke in an interview: "When they opened the Great Pyramid, in the king's chamber they found it totally empty except for this white powder all over. I think it is just the material that was being used in there that was spilled. Um you know the Great Pyramid is quite a mystery to most people, why it was built, why it's the way it is. But, ah, the white powder, I would give anything to have just a few milligrams of the white powder that actually we know came out of the Great Pyramid. Because it will analyze to be alumna or silica it's the only thing it will analyze to be, and so you'll assume there's nothing to it, and you just dispose of it. But it's not alumina or silica...

I think it is Manna. I haven't got any to analyze, but I can tell you that the Vibuti, the Sai Baba manifest, in fact is iridium in the high-spin state. That we have analyzed it and he does ingest this material. Iridium polymerizes with hydrochloric acid, in the stomach you have hydrochloric acid and so when you take it in high amounts in your stomach and then he regurgitates this, and they call it Lingam..." "Well they worship it as the holy material, and he literally eats it in macro amounts. His body it totally saturated, and they claim he can bio-locate."

The only inscription inside the great pyramid is the Egyptian hieroglyphic symbol for bread. Here is this giant building, which our eminent scientists cannot figure out why it was built, how it was built, and what it was used for and the only Egyptian inscription inside the whole building is, "bread." The meaning seem to go over the top of their heads. The actual translation of the bread symbol is: "What is it?" This is also the meaning of the word 'manna'. A poem in the ancient Egyptian Book Of The Dead ends each line with the question: "What is it?" They used this same hieroglyphic to refer to the mysterious white powder of gold. Ask any Rabbi about the manna or white-powder-of-gold and they will say: 'Yes we know about it, but we no longer know how to make it.'

When Adam and Eve were given the ability to procreate, it was essential to bar access to the "tree of life" by which they might live forever—otherwise there would have been an uncontrollable population explosion. According to the Bible Adam lived 930 years.

Mankind was genetically manipulated; created from that which is from the earth or earthling and Angel DNA. Then some space virus infected all of mankind contaminating the genes so that we now live to the maximum age of 120 years. Mankind was supposed to be a perfect being and never get sick. The gods were going to kill us all off but then decided to let this batch live anyway to see how it would turn out.

HOW I DISCOVERED IT

I always had an interest in gold prospecting and I studied the formation of gold deposits more than twenty years. I have taken several geology courses in college and done hundreds of mineral tests in the laboratory. I conferred with many gold miners in California and Alaska. My father owned a gold mine south of Fairbanks, Alaska on Little Moose Creek. I built his mining equipment because I weld and fabricate steel. I also have a background in electrical engineering. From this basic knowledge and practical experience I was able to put together details from several writers and scientists that enabled me to figure out how the Ark works.

It all started one day as I was walking up the beach after a days work-picking salmon out of my nets. I noticed strips of heavy, black-sand on the beach. Ten years later I still wondered what this sand was composed of. Finally, after twenty years went by I sent two samples into a laboratory in California requesting a spectrographic analysis and a leach assay. The results were puzzling. The leach test showed little or no precious metals. The spectrographic analysis showed some copper and zinc and three hundred pounds of iron per ton and no precious metals.

Finally I totaled up all the weights per ton of all the elements in the test and there was six hundred pounds missing. Where did the other stuff go and what was it? Where did it go? An analysis is supposed to show everything that is in the sample, right? Many more years went by until 1998 when I read about monatomic minerals in Nexus magazine. Orbitally Rearranged Monatomic Minerals are the most astounding thing. Plants readily extract this stuff from the soil and it is necessary for cell division in all-living things. You eat this material every day in your food and you don't even know it exists. Hardly anyone is aware that this material exists yet it is the most nourishing food on Earth. You don't get enough of it in your diet because you eat food that has been grown on the same soil for over a hundred years. Some farms have been growing food on the same soil even longer than a hundred years and the minerals are so depleted that the food grown in this soil is worthless.

Taking this one step further you could say that America has become a nation of zombies who think they are getting good nutrition when in fact their body cells are starving. Is it any wonder why so many chronic diseases are on the rise in America today? Much like past civilizations we may be killing ourselves with our diet.

Over the years I figured out how the black sand where I live in Tuxedni Bay is concentrated from the volcanic mud. Subsequent eruptions of Mt. Redoubt deposited billions of tons of volcanic mud forming the twenty-plus, square-mile, area where I live. I didn't learn this fact until a couple years after Mt. St. Hellens erupted in Washington State. The federal park ranger for Lake Clark Park explained the process to me. At the time of the ancient eruption the volcanic mud flowed several miles out into Cook Inlet. During the next nine hundred years or so several million cubic yards of this mud and boulders were washed away by ocean waves. The southeast winds pile the heavy sand and gravel up onto the beach. Most of the lighter material was washed away leaving big boulders perched here and there. The south-west winds pile the heavy sand higher on the beach while carrying the lighter sand up into the grass inland adding to the peninsulas and forming sand spits. It took me years of observing nature to figure this out.

Spectrographic tests of the black sand still puzzled me. Each time I added up all the pounds of elements per ton there was still over 600 pounds missing. What was it and where did it go? After many years or reading California Mining Journal I saw an advertisement where they test samples with electron beams.

I didn't know it at the time but over half of the periodic table of elements can exist in the ORME or m-state lacking electrons. ORME atoms share a double number of protons and neutrons in the nucleus. The electrons have to orbit out eight times further from the nucleus in opposing directions to balance the atomic forces of the Hugh nucleus. The electrons can sometimes exceed the speed of light reflecting pure light back leaving it a palpable white super-conducting powder with thousands of amps of electricity flowing through it. Platinum and gold cannot form a metal when in this state because they lack electrons to from a bond. You see, everything on Earth is held together with electrons. Every atom in your body and everything you perceive to be real is bonded with electrons.

I sent a sample into Bremner Technologies of Las Vegas. The have a machine with three big electron guns that melt samples in an argon filled chamber. This process imparts electrons which substantially increasing the precious metal values of the sample. After the lump of molten glass cools they break it up and leach out the metals just like any other assay office. The gold and platinum values were quite promising. I realized at the time that the technology of recovering precious metals using the electron beams was a little beyond my means.

Several more years went by then I read David Hudson's lectures in Nexus magazine. I accidentally picked up a copy of Nexus magazine in a friends office while I was in San Diego. These three lecture transcripts changed my life. All of a sudden all the unknowns and questions I had about make-up of matter were clear to me. It explained everything. It explained why humans had psychic abilities. It explained why the hundred or more people mentioned in the book of Genesis lived over nine hundred years. It explained why the pyramids were built and with a little reading from other sources I figured out how the Ark works. It explained how deadly diseases are cured genetically. Now I knew what the missing stuff was in my beach sands. It was platinum and gold in the ORME state.

David Hudson's interview: "Remember they said in the Bible that the Ark of the Covenant actually levitated and floated along and actually carried some of the people who were carrying it. The only thing to do that is a superconductor." It was creating an antigravity field.

There is enough raw material in the sands and volcanic ash of Tuxedni Bay to manufacture the anti-gravity gold of the Gods forever. I see no reason why I can't distribute it to the world via the Internet. This could be a multi-billion-dollar-business, which helps people live longer, healthier lives. It enables your body to cure itself of cancer by correcting your DNA. It cures hundreds of diseases by strengthening the immune system.

Most of the people mentioned in the book of Genesis lived over nine hundred years. According to the Bible Noah lived nine hundred and fifty-eight years. Many of us don't read the Bible because it was force upon us when we were kids. I read the Bible because it is the written history of mankind.

I read it from the standpoint of trying to decipher the cryptic meanings hidden within. According to Genesis chapter 5 lines 5, Adam lived 930 years. His first son, Seth lived 807 years. How could men live that long? When I was a kid I didn't believe it was possible for a person to live nine hundred years but after reading about ORME or m-state stealth-atoms and the white-powder-gold I now believe that most of the Bible is true. Enclosed are phone numbers and addresses of references that document what I write here. I strongly suggest that you call these numbers and read the material to verify what I am telling you. Check out David Hudson's patents on Orbitally Rearranged Monatomic Elements.

NATIONAL PUBLIC RADIO NEWS (MARCH 25th, 2001)

"Scientists are finding out that clowning is harder to do than expected. Random DNA errors occur producing animals that are susceptible to disease. If the first steps of cell division do not go according to plan the organism has many genetic defects. It all takes place in the first few minutes of cell division." The missing ingredient of course is the super-conducting, white-powder-of-gold.

According to the fossil record modern man or Cro-Magnon man suddenly appeared on Earth approximately 300,000 years ago with no apparent evolutionary record leading up to his appearance. The heavy-set Neanderthal man and Homo Robustus have fossil records dating back millions of years. This creation argument agrees with the Bible as well as ancient Sumarian texts and it explains the fossil record or lack thereof. My argument covers three of the bases but it is still not a home run.

Much like the offspring of a male ass and a mare, which produces a mule that cannot reproduce, the genes of Cro-Magnon man were engineered so that he could not reproduce with either the Neanderthal or Homo Robustus thereby ensuring that the genetic lineage would not regress but continue on as a new species. God wanted companions not supplicants. Adam was a blue blood, a type Y blood with a zinc base. Jesus was also a blue blood and so were most of the kings of Europe and England. The kingship lineage was passed down from generation to generation. It is very important that Man does not clone himself because he does not know what he is doing. Just because science has mapped the gene code does not mean they understand it. If they should accidentally produce a new kind of human and that human mated with other humans it could destroy what the Anunnaki Gods have been trying to do for thousands of years and we may find ourselves hunting with rocks and sticks again. The gene code was much simpler than expected with only thirty-eight genes. Scientists are now saying that where the real magic occurs in the human Genome in the interaction between the genes. If this is true then all the study and effort are up in smoke. If the Anunnaki Gods were like the little Gray Aliens that have been visiting Earth for such a very long time, manipulating the DNA of both plants and animals, then new evidence shows that the Anunnaki could very well be related to both plants and animals. This is a startling revelation!

GORDEN MICHAEL SCALLION

In the February 14, 2001 broadcast of the Art Bell Show, Gordon Michael Scallion reported that he was having a part of a Gray Alien fingernail analyzed and it looks to be related to both plant and animal cells. "Unlike animal cells which have flexible cell walls plants cells have square cell-walls made our of cellulose fiber. The Alien cells have a similar structure to both plants and animals."

Could it be that they use their own body cells to create both plants and animals so very long ago---possibly billions of years ago? Any beings that can live 350,000 years like the little gray aliens would have an entirely different outlook on life than humans who have difficulty planning ahead one year. Development of a new species of man even if it took as much as 30,000 years is not that much time for someone who lives 350,000 years or longer.

In the March-April 2001 issue of Nexus Magazine Richard J. Broylan, Ph.D. reported an incoming object. "There is a large incoming object expected, but it is quite a different kind and it is due around 2011-2012. Navel observers have already spotted it, coming from a direction visible from the southern hemisphere. Z confirmed that this is the 10th planet of our solar system, which takes an extremely elliptical orbit. He said it is the one which orientalist scholar Zecharia Sitchin wrote about after decoding the Sumarian tablets. These tablets spoke of a 10th planet with a 3,600-year orbit cycle. That planet, which they call Nibiru, is inhabited by the extraterrestrial who told the Sumarians that they bio-engineered humans using Earth primate stock and extraterrestrial genes."

THE FACAD

Dr. Michael Heisenburg of University of Wisconsin, Madison was on the Art Bell radio show February 23, 2001. Dr. Heizer wrote a fiction book called the Facade. Dr. Heizer is a great deal more knowledgeable about Hebrew and other ancient languages than I. Dr. Heizer made many interesting discoveries about angels and miss-translated words in the Bible.

I find it incredibly synchronistic that Dr. Heizer would be on the radio one day after I wrote the above account on how God created man. He was scheduled to come on the show a week in the future but Art Bell had a cold and Ian Punnet was filling in. The person that Art Bell was going to interview dropped out so I got to hear Dr. Heizer. His web site is: www.facadnovel.com.

"Who are the Angelic binges mentioned in Genesis? Where they Heavenly beings or extraterrestrial? The humans that were created like Adam were half human and half angel or in the Hebrew language Melakheim. Are we fallen angels? Who were Michael, and Gabriel? Ural, and Metatron? What are their hierarchal relationships? These transcendental beings existed on many levels as inter-dimensional entities.

What are the links between the Hebrews and gray aliens? The angels did come from heaven but were not acting in accord with the Divine Council when they took up with the human females. -Genesis 6 Two angels were left behind to guard the gates to Eden when Adam and Eve were expelled. These don't sound like angelic beings to me.

The Divine Council is a hierarchical set up of divine beings or Elokheims. The word Eloheim (Gods) is a plural word. Somehow it got turned around into a singular-one God during the translation of the Bible for the Christian Religion.

Below the Eloheim are many lower ranking Angels. Next there are the Malaheim, Saraheim or Seraphim, and Cherubim or Cherub. Whenever you have one of the high council members come to Earth and interact with men you can't tell them from men. They look just like men. The required dress when coming to Earth is human flesh. Yahweh is the chairman of the High Council.

When the Serpent tells Eve to eat of the forbidden fruit of knowledge of good and evil and her eyes will be opened why isn't she shocked when the serpent, Nakosh talks? Nakosh is translated as the shiny one, 'the bronze or brazen', son of God. The reason she isn't shocked is because Eden was where the high Council met and Eve was very used to these beings coming and going every day. When Nakosh spoke she was not alarmed. Eden can be translated as: "mountain." Eden was a mountain where the Gods mined gold. The reason I know this to be true is because I am a gold prospector myself. What else would someone be doing up on a mountain?

This incident where Adam and Eve were kicked out of Eden causes a confrontation between the Gods of the high council. Nakosh is made to crawl along the ground the rest of his life. I can't get over the similarly between Nakosh and Quetzalcoatl. In Aztec art he is depicted as a feathered serpent. On a pyramid dedicated to him the setting sun illuminates the right-angel stones blocks down the corner of the pyramid making it look like a wavy snake. A snakehead is carved in the bottom corner stone of the pyramid. On certain days of the year you can see his likeness. To the Aztecs he was a God who came down from heaven in the shape of a snake and taught them their science and technology.

Nakosh, (E.A.) the serpent is described in the Qumran Cave Dead Sea Schrolls as a watcher or fearsome like a serpent. YHWH punished Nakosh for corrupting the humans and uttering the statement: "I will be higher than the most high. I will be like the sons of God." Obviously he was not one of the sons of God he was a God. YHWH punished him for his arrogance by turning him into a reptile. The Divine Kings are a cross between a divine being and woman. The product of these unions is called Nephalene or fallen angels. These were the "heroes of old, warriors of valor."

After Genesis 6 where the gods took the human females as they pleased God has to make a decision because the human line was corrupted. God decided to scrap everything and started over after the flood.

God divided up the nations according to the number of the sons of Israel. Another translation is the number of the sons of God or 70 nations. God says; "OK if you persist in following these lower deities go ahead and worship Vishnu, Baal, Bel, Ashtoreth, and all the rest." The lesser gods then traveled to all parts of the World to colonize it with humans thus forming the various races of people.

The God Enki, or Thoth, also known as YHWH went and started over forming a new nation out of the seed of Abraham. The book of Enoc and at least twelve other books dealing with the history of the Hebrews were edited out of the Bible by the Christians so that they could create their own religion. This period of history deals with the time immediately after the flood. The people who believe in the one God, (monotheists) have been warring with the people who believe in many Gods ever since.

The Greeks had the Titans, sons of God who came down from heaven to do battle on Earth. The word Titaniss in Greek literally means gray. So the sons of God were gray aliens. The Facad is a cover story; what they tell mankind to cover up all these secrets.

GOLD MINERS

Every account I ever read about Gods, more often than not they are living up on a mountain. The only thing productive one can do up on a mountain is mine gold. You could say they were admiring the view but that would only take five minutes. After that they would have to go back to mining gold. Gold was the real reason they were here in the first place. What else would they be doing up on a mountain? They were gathering and manufacturing the Anunnaki Gold and transforming it into manna to eat.

ANCIENT AMERICA

Gold mining was an important occupation for many of the ancient Hebrew. One lost tribe of Israel wound up in the Americas in what is now the state of Louisiana. From there they migrated south to Mexico City then for some unknown reason traveled back up the Mississippi River to the part of the country known as Illinois. We know this because archeologists found inside of one of the large burial mounds a clay tablet written in ancient Hebrew.

In the state of Michigan a farmer found a stone tablet in the 1800's. He kept it for some time---finally sending it to the Smithsonian. It laid around the Smithsonian several decades because nobody could read it until some time after 1910 a New Zealander, Harvard, college-professor by the name of Berry Fell identified it as Hittite. After that the Smithsonian sent it back to the farmer who gave it to them and it was probably destroyed. What is interesting about this is the stone was discovered in the 1800's before the Hittite language was translated. This rules out the possibility of the artifact being a fake. The Hittite language was first translated in 1910.

Some unknown people mined 30 million pounds of copper in Northern Michigan about 10,000 years ago and about the same time the people of Mediterranean were having a copper age. Why can't mainstream science put it together?

GENETIC TESTING

Genetic testing is still a relatively young science. This branch of science is wholly based on testing the DNA of the mitochondria. This is a microscopic bacterial organism that your body cells use to metabolize food. This organism is not a human organism however. At some remote time in the past it formed a symbiotic relationship with human cells taking over the digestion and production of energy. This symbiotic relationship took place a very long time ago, perhaps billions of years. It is the heating-oil distributor of your body.

The method used to trace mitochondria DNA is based on the fact that it is only passed down from the female. It is theoretically possible to trace this DNA back to the original female (EVE) who created the modern race of people who now inhabit earth. The process involves extracting the mitochondria from blood or body cells. So far geneticists have identified forty-two types of mitochondria DNA. That is all they have found so far in all the human species of the world. Now I believe the number of DNA types is over fifty.

The method used to categorize DNA is a letter and number system. DNA types are designated as type A, B, C, D, etc. Then when they ran out of letters in the alphabet they started over with A sub 1, etc. Halpro group-X DNA however is the DNA of the Hebrew people. Of course it is found in Israel but it is also found in the Gobi Desert and can be traced back 10,000 years when these people migrated from Atlantis. Halpro group-X can also be found in the Iroquoi Indians. Most all other American Indians have DNA similar to the people living in Siberia, which does not prove however that they walked over here on an extinct land bridge.

MOUNTAIN OF GOD

According to the Bible we know that God lived on Mt. Sini. We know that there are satellite photos that show caravan routs leading to Mt. Sini (Jabal Al Lawz). It would be the ultimate archeological discovery if someone were to actually find the mountain of God. The knowledge that the translation of word Eden means mountain opens up a whole new field of archeology. I hope scientists start studying satellite photographs of caravan routs leading to mountains. They will either discover gold or the mountain of God or both.

The reason why I think along these lines is because I am a gold prospector. The life and habits of ancient civilizations and Gods seem to puzzle scientists but to someone with a little practical experience of survival in the wilderness it all seems so simple. Too much education seems to befuddle their minds. I would love to go to Mt. Sini and look for God's mountain. What could compare with that? If someone finds God's Eden Mountain after reading this book I want to know about it. It would make me very proud.

A SINAI GOLD MINE

Has anyone dated the charcoal or carbon near the forge? When it was excavated in 1948 they didn't have carbon 14 dating. If it turns out to be 3,500 years old it could have been left by Moses goldsmith, Bazalele. It could be much older if wasn't used when Moses was there. If it turned out to be 30,000 years it could be Eden. Carbon 14 has a half-life of 6000 years. If you have a hundred carbon14 atoms in your hand, six thousand years from now you will have fifty. Counting the atoms in subsistence can be used to determine it's age if it has not been contaminated by burning or extensive radiation. We must find a way to accurately date this site. If Mt. Sini (Eden and Jabal Al Lawz) turned out to be 30,000 or 3 million years at least we will have an idea of the time scale.

In Howard Blum's book: Gold of Exodus When Bob Cornuk and Larry Williams were exploring Jabal Al Laws, the mountain of God, they found a cave on the opposite side of the mountain to where the Saudis were building a radar tower. It was getting late and Bob wanted to go inside but Larry Williams refused because he sensed a presence. The two men spent the night sleeping fitfully outside the entrance of the cave.

Boeing Aircraft of Seattle, Washington had the contract to build the radar tower. One their employees, a Philippine construction worker took a walk during his lunch break. He discovered a cave and went inside. He turned up missing for three days and was found wandering around on the mountain in a sort of trance. He reported that he had entered a cave and became immobilized by a spirit. He reported that the spirit held him captive and sung to him. He said it was torture. All he could think of was getting away. Was this Thoth or Enki / YHWH? Whoever it was sought company and simply wanted someone to sing to.

CHAPTER TWELVE

DAVID HUDSON

David Radius Hudson is credited by most people in the field as being the originator of the term "Orbitally Rearranged Monoatomic Elements", or by its acronym, the ORME.

Mr. Hudson began speaking publicly about his research and discoveries in 1995, when at that time he pointed out the multiple connections between the works of Zecharia Sitchin [1], the Anunnaki, the Tree of Life, The Egyptian Book of the Dead, Alchemy, Immanuel Velikovsky, Superconductivity, Ark of the Covenant, The Adam's Family, and more recently the works of Laurence Gardner. [2] There are also numerous researchers interested in a variation on the ORME, the ORMUS, where it is assumed that the working material is not necessary mono-atomic, but may be diatomic—consisting of two (or more) atoms of the precious metals.

The story of Mr. Hudson's efforts over the years is given in the form of a rough transcript of his presentation in 1995 at the International Forum on New Science in Fort Collins, Colorado. As such, it serves as an excellent introduction to the study of the ORME, and its immense implications. His story and the profound revelations of his work are well worth considerable study. This is only the beginning introduction.

DAVID HUDSON AT IFNS

I was buying gold and silver as an inflation hedge. Then got into producing gold from a natural source, old mining sources. Mining worked well with the farming. You beat Uncle Sam as much as possible out of taxes, and at the same time accumulate as much wealth as you can. Leaching gold with cyanide process is like leaching salt out in the farming process. More of a hobby than a business—no intention of making money. But something I enjoyed very much. Did it for fun.

In the process of recovering gold and silver, I began to recover something else, which was causing losses of the gold and silver. Eventually, it reached the point where the gold and silver would not recover at all because of the something else. I then shut it down, to find out what the problem material was. I am not a physicist or a chemist and had no idea what the stuff was. It would recover and had a specific gravity; it would recover in the molten lead like it was gold and silver; it would flow out of the lead; but when I held the lead down, I had nothing. The mining community refers to this as "ghost gold", a non-assayable, non-identifiable form of gold.

I then became involved with someone who does emission spectroscopy (ES) and became aware of work done by the Soviet Academy of Sciences. When one does ES, it involves taking a carbon electrode, placing your sample on the carbon electrode, and then running a second carbon electrode down above it, and striking an arc. When you strike the arc, the elements ionize and give off specific light frequencies. This is spectroscopic analysis. In the analysis, it's done for 10-15 seconds before the electrode is burned away, and American Spectroscopists claim that anything there will be ionized and read within those 15 minutes.

My sample was identified as Iron, Silicon, and Aluminum. I then spent three years in finding ways to take away all the Fe, Si, and Al. Then, I still had 98% of the sample of the material. On the arc, the material didn't indicate to be anything. It was nothing. Back to Cornell University, where I worked with a Ph.D., who did X-Ray Analysis. This involved: Cumming Microscopy, Diffraction Microscopy, Fluorescent Microscopy, and five other wonderful technologies. The Ph.D. said that it was Iron, Silicon, and Aluminum. Stayed there to remove it all. After that, the Ph.D. said it was "pure nothing." This wasn't good enough for me. I could hold it in my hands, weigh it, perform chemistries with it—it was something. I then recalled that, according to the Soviet Academy of Sciences, the proper analytical tool is to burn the sample in the emission spectroscopic analysis for 300 seconds, not just 15.

To do this, you have to sheath the electrode with an inert gas—keeping all the oxygen away from the DC arc. Otherwise, the carbon oxidizes, and the electrode falls apart. I set up to do this, using Argon gas to sheath the electrode—keeping the oxygen away. Because carbon is a very high-temperature material, it will then last for 300 seconds. When the material was placed on the electrode and the arc was struck, there was no reading at all for 15 seconds [other than "electronic grass" on the scope, as well as Iron, Silicon, Aluminum, and occasionally, traces of Calcium]. Then the material went quiet.

[Argon gas if fundamentally crucial to Sonoluminescence, as well. It also has an identical crystalline structure to such elements as Rhodium and Iridium.]

Finally, after 90 seconds, Palladium (Pd) began to read; after 110 seconds, Platinum (Pt) began to read; at 130 seconds, Ruthenium (Ru); at about 140-150 seconds, Rhodium; at 190 seconds, Iridium; at 220 seconds, Osmium began to read. The Russians call this fractional vaporization. For example, when one has water in an iron container, you can't get the iron hotter than the boiling point of water as long as there is water present. This is the basis for cooling systems in engines, and why automobile engines don't overheat, as long as there is water present. But once, the water is gone, the temperature rises very rapidly to the melting temperature of the iron.

[The temperature of any well-mixed solution undergoing a phase change will stay at the temperature of the phase change until the phase is completed. Also, the relevant boiling temperatures for the metals in the sample are: Calcium: 1420 oC, Iron: 1535 oC, Silicon: 2355 oC, Aluminum: 2327 oC; followed by: Palladium: >2200 oC, Rhodium: 2500 oC, Ruthenium: 4150 oC, Platinum: 4300 oC, Iridium: >4800 oC, Osmium: >5300 oC. (Silver has a boiling point of 1950 oC, while Gold's boiling point is 2600 oC.)]

Essentially, all of the emissions from the elements were coming off in the sequence of their increasing boiling temperatures. The maximum temperature of the DC arc is, theoretically, in the center of the arc, 5450 to 5500 oC; while the sample was slightly away from the center. Thus, all the heat went into boiling off one element at a time, in the sequence of their boiling temperatures. They come off individually as if there is nothing else in the sample.

I continued to run the sample for 2 and a half years, comparing it to standard samples. The amazing thing is that commercially available samples of the precious metals, when placed in the emission spectroscopic DC arc, read within 15 seconds. (And they assume they're reading it all.)

But then it goes quiet, until after 90 seconds, it starts to read again. About 85% of the reading occurs at the end. In effect, the people buying the commercially available samples and doing readings, are only doing about 15 to 20% of the sample. And they assume it's everything. Short burn times don't do the trick. They assume the standard, and yet this is not the correct standard.

Keep in mind that the Soviet Academy of Sciences, the most prestigious scientific body in the Soviet Union and Johnson-Mathewe-Inglehart produce all the precious metals in the world. The mining activity of the best deposit in the world in South Africa for six of these elements (Pd, Pt, Os, Ru, Ir, and Rh—i.e. no silver or gold) may yield only one-third of one ounce of the precious metals per ton of ore. They go a half mile down into the ground, following an 18-inch seam of material, to get 1/3 of 1 oz per ton of all the precious elements. No one else knows it's there, and no one can analyze it. We, on the other hand, can derive and identify out of one ton of ore: 6-8 ounces of Palladium, 12-13 ounces of Platinum, 150 ounces of Osmium, 250 ounces of Ruthenium, 600 ounces of Iridium, and 1200 ounces of Rhodium!!

This was then confirmed by a highly respected chemist and spectroscopy, including all of the colors of the solutions being correct, all the oxidation potentials were correct, all of the physical properties were correct. An analytical chemist. Consider Rhodium. Rhodium produces a crimson, blood red colored salt. That is how it got its name, from the rose-colored salt, and the only element which produces this color. Very conspicuous.

When you precipitate Rhodium out of solution, you add bromide as the oxidizer, and then you do a neutralization of the acid, and the hydroxide precipitates out of the solution. You filter it, dry it, oxidize it, hydrogen reduce it, and you should have metal. (Standard procedure). But we neutralized the solution of the pure Rhodium, got a red brown dioxide, filtered that out, dried it, and heated it in a tube furnace under oxygen up to 850 degrees for an hour to dehydrate it, and we'd get this red brown dioxide. Then we put it back in the tube furnace and hydrogen reduce it to a gray powder, and then take the gray powder in a tube furnace at 1000 degrees under argon, and it turned snow white.

A commercial spectroscopic firm then analyzed three samples, and again picked up Iron, Aluminum, Silicon, and Calcium. There was no consistency between the three samples, which were all the same. The material was 99.9% pure Rhodium, in different stages of the processing.

The standards that are sold as labeled as $RhCl_3$, when in reality they are $Rh_{12}Cl_{36}$. It still has metal-metal bonds. Even without the Chlorine, you still have the metal bonds, which are never lost. But if you take Rhodium to the monatomic state, you can end up with $HRhCl_4$. Then when you take away the Chloride, you get HRh (Hydrogen Rhodide). A Rhodide is a -1, instead of a +1. The physical properties are more like an Iodide.

When gold is produced as a monatomic gold, it's a forest green color. As a metal, it's a yellow gold. No one has monatomic gold as a commercial product. Monatomic gold is much more powerful, as in a fuel cell. Boiling gold will never result in a monatomic gold. Gold has the 5d, 6s1 electron structure (the single s electron, like Sodium, Potassium, Hydrogen, and Lithium), and is thus explosively reactive. Except that in the case of gold, it's gold reacting with gold.

But in the bowels of the earth, in the volcanoes, nature is producing monatomic gold. When it comes out, 98% of the gold coming out is monoatomic, 2% is metal. [Thus, Hudson may not be making monatomic gold, he is separating it instead.] We have worked with the yellow gold, converting it, but always coming back to yellow gold. But when we get monatomic gold, it never goes back to yellow. And as monatomic gold, it is not metallic, has none of the metallic qualities of yellow gold.

If you use thermo-gravimetric analysis, and you produce monatomic gold, you get sort of a gray-black, Hydrogen Auride (HAu). When you heat it, and the proton is annealed away, this is the same way you produce Amorphous Silicon (Silane to Amorphous Silicon). When you heat it, the proton is annealed away, it goes to a snow white powder! It loses 4/9th of its weight. If you take it back to metal, it regains the weight. As you kept annealing the material, it would levitate – taking the pan with it. In cooling, it would sometimes go to 2 or 300% of the weight. In heating, it goes to less than nothing.

This only happens in the white powder form. But mass has never left. Losing weight when cooling the material (approaching absolute zero), and you have a superconductor.

A superconductor is a material that has a single wavelength in the sample, a single vibration or frequency, much like a laser. By definition, a superconductor does not allow any voltage potential to exist within it; it's perfect amperage, but no voltage. To hook up wires with ordinary current to the superconductor and get the electrons off the wire, you need voltage. You have the tune the vibrational frequency of the electrons in the wire to that of the superconductors. And to get them off.

The electrons going into the superconductor have to pair, with a time forward electron with a time reverse electron. When they pair, they become light. Any amount of light can exist in the superconductor. It doesn't reside in any space-time. The only way to prove it's a superconductor is to measure a Meissner field. Non-polar field (only field of its kind). Superconductivity responds to magnetic fields. Earth's magnetic field is larger.

A superconductor can see your thoughts in your brain. Different parts of your brain light up when you eat something sweet or something sour—it's a superconductor that sees it.

When it goes to the white powder and loses 4/9th of its weight, it's flowing light within it, in response to the earth's magnetic field. And it flows so much current that it levitates 4/9th of its weight on the earth's magnetic field. A human hand has sufficient amperage, that if you pass it under the sample, the material floats. It's that sensitive to magnetic fields. All the eight precious metal elements can do this. Also Copper, Cobalt, and Nickel. So, I filed 11 patents.

In 1990, my uncle showed me the Time-Life Book, Secrets of the Alchemist. I was not interested in Alchemy; I wanted credibility in physics and chemistry. The book talks about a "white powder of gold". The goal of the alchemist had been to make a "white powder of gold". To make "the container of the light of life." If you stand in its presence, you don't age. If you partake of it, you live forever. I begrudgingly agreed to read the book I have since read some 500 books on alchemy, alchemy history, and it all goes back to a man named Enoch. Thoth, Hermes, Trigeminus. Same man. Ascended by partaking of the white drops. He never died, instead ascended because he was so perfect.

I found a huge amount of research going on in treating cancer with precious elements. That the elements have been found interacting with a cell by a vibrational frequency or by a light transfer and have been correcting the DNA. Any alteration, any defect in the DNA is corrected by the precious elements. It perfects the cells of our bodies. But the element going into our bodies is not a metal, the element is not a heavy metal, it is an element. So there's no heavy metal poisoning. You can eat any amount of this white flour, and it doesn't hurt you. If you eat it, it just goes through your digestive system.

I took some brain tissue from a pig, and some from a cow, and analyzed it. They destroyed the organic and did a metals analysis. Over 5% of the brain tissue by dry matter weight was Rhodium and Iridium. But no one knows it because it can't be directly measured. The elements are flowing the light of life in your body. The elements are in fact what the light is. There are four papers by the U.S. Naval Research that they have proved the cells communicate with each other by a process identical to superconductivity. But they can't figure out the physical mechanism.

It is the stealth atoms. It's the atoms in our bodies. However, no one knows they're there because they don't identify them by instrumental analysis. And the reason they don't identify them is also in the literature. Since 1986, the top physicists in the world, at the Niels Bohr Institute, at Argonne National Laboratory, at Brookhaven... They have found that there is a group of elements in the center of the Periodic Table that go through this strange state of existence.

So most of these publications occurred between 1988 on, but my patents were filed first. What they have found is that the nucleus of these elements were deformed, went to a high spin state—what's called high spin nuclei—and theoretically, these high spin nuclei should be superconductors because high spin nuclei pass energy from one atom to the next with no loss of energy.

This is what is in our bodies. This is what flows the light of life. And when you understand that a superconductor flows a single wavelength of light, but, in fact, the light is a null light, two ways that are mirror images of each other. There's no wave—it appears to cancel, but the null wave is in fact, while not measurable directly, is what produces the aura around our bodies. The aura is the Meissner field of superconductivity.

In our bodies, we have all this junk DNA. There are 30 aspects of the DNA that nobody can figure out what it's there for. We only use 15% of our brain. What's the other 85% of it there for? Did we evolve a brain we don't use? It's as if we had, at one time, a higher state of enlightenment, and we have fallen into the state we exist in now.

In ancient Egypt, which I traced this back to, there's a book, called The Egyptian Book of the Dead and the Papyrus of Ani, by Budge. This is the oldest book of the dead, from Old Kingdom Egypt. They found it, dating from about 3500 B.C.E., in the tomb of Pepi II. It says, "I am purified of all imperfections. What is it? I ascend like the golden hawk of Horus. What is it? I pass by the immortals without dying. What is it? I come before my father in Heaven. What is it?"

It goes on and on and on. It keeps asking this question, "What is it?"

The Hebrews worked in Egypt for many, many generations—they were the artisans and the metallurgists. When they left Egypt, Baalzelael, the goldsmith and Moses prepared the bread of the presence of God. He prepared the bread that the high priest partook of, the Melchizedek priest. The word in Hebrew that means "What is it?" is manna. The word, manna, literally translates, verbatim, into a question, "What is it?" If you don't believe it, look in the Travels of Josephus. The very same words that were used in Old Kingdom Egypt, 3500 B.C. [Pepi II reigned 90 years from c. 2300 to 2210 (traditional dating), or 1720 to 1630 (Immanuel Velikovsky's dating).]

These elements are naturally in your body. It's primarily Rhodium and Iridium. Now the Bible says that Moses told the Hebrew people, "You have not kept the covenant, and so the manna is being taken from you. But it will come back in the end times. When we will be a nation of high priests, not an elect high priesthood." This is the food, the light, you take in your body.

If you ask a Rabbi, have you ever heard of the white powder of gold, he'll say yes, we've heard of it, but to our knowledge, no one has known how to make it since the destruction of the First Temple. The temple of Solomon. This knowledge has been lost. But it wasn't completely lost, the high priests when they left the temple when it was destroyed, went out on the desert and organized a commune called Qumran. They were the Essenes. In The Dead Seas Scrolls Uncovered [Eisement and Wise] this in ancient times, when the white powder was mixed in water, was known as The Golden Tear from the Eye of Horus. It was called, "That which issues from the mouth of the creator." The spittle. Not the word of God, but the spittle. Or the semen of the father in heaven. If you put the white powder in water, it doesn't dissolve. It forms this gelatinous suspension and looks just like a vial of semen. Being a farmer, I know what semen looks like.

The symbolism of "prepare yourself like a bride in the bridal chamber", purify and cleanse yourself, prepare yourself for the coming of the Father in heaven, to be inseminated by the father in heaven in the bridal chamber. To totally be regenerated, to be purified, to be cleansed. Every cell in your body will be taken back to the way it's supposed to be, when you were a teenager or a child. It perfects the DNA. And it flows the light until you reach the point where the light body exceeds the physical body.

In ancient Egypt, they said you have a physical body you must feed to grow as it's meant to be. If you don't feed that child, he'll never grow. He'll never become the person he's supposed to be. But you also have to feed the spirit body, you have to feed the ka, what they called it in ancient Egypt—so it can grow and become what it's meant to be. And most of you aren't feeding your ka. It's sitting there like a little runt inside your body. And they said you feed it and feed it and feed it with the semen of the father in heaven, and it grows and grows and becomes more enlightened and more enlightened, and you reach the point where the light body exceeds the physical body. You light up the room when you walk in.

The gifts that go with this are: perfect telepathy, you can know good and evil when it's in the room with you, you also can project your thoughts into someone else's mind. You can levitate; you can walk on water because it's flowing so much light within you that you don't attract gravity. And when you understand that when you exclude all external magnetic fields, when you exclude gravity, you are no longer of this space-time. You become a fifth-dimensional being. You can think where you would like to be, and go there. Just disappear. You also have other attributes that they go into. You can heal by the laying on of hands, and can cleanse and resurrect the dead within two or three days after they died. You have so much energy that you can embrace people and bring light and energy back into them.

Sounds pretty far out. Most groups don't receive this very well.

If this is really what this stuff does, then let's use it. I haven't achieved everything yet, but it miraculously has cured every disease that we've tested on thus far. Started with very incremental amounts of 2 mg (32,000 mg in an ounce), and have gone up to 50 mg—50 mg over a period of 60 to 90 days, cures cancer, AIDS. It's the light that corrects itself. You all know this.

Christ said to his disciples, "Don't touch me, I don't have on my earthly garments." They asked, "When will we see you again?" He replied, "When you have prepared the proper food and have on your proper garments." What is the proper food? It's the food of the angels, the food of the gods, the manna, the "What is it?" And your proper garment is your garment of Or, your Meissner field (what science calls it). And literally, it's about a thousand times what you have now.

The amazing thing about superconductors is that they don't have to touch for their energy to flow from one superconductor to another. Electricity has to touch. Superconductors can sit at a distance from each other, as long as they are in resonant frequency with each other, they are One. They function as one. So when you have your perfect superconducting body, you're not of this space-time. You are a light being, and your mind is one with other people's minds. You literally know their thoughts, and they know your thoughts. You and they are literally of one mind, one heart, and this is science.

The Bible says that the man who will plant the golden tree of life, which in Hebrew is the ORME tree (the name of my patents—at the time I had no idea of the connection). When my cousin joined the Morman Church, she had to do their genealogy, and his great, great, great grandmother is Hanah DeVries, daughter of Christopher DeVries, brother of Claude DeVries [see reference 2 above, and Holy Blood, Holy Grail]. Nostradamus worked for the DeVries family, and Nostradamus prophesied that by 1999, the occult gold will be known to science.

The old enemy of religion, the old enemy of philosophy is science, but in fact, science will serve up the confirmation and science will be the one who brings this to the world.

Religion has tried to do this for two thousand years, and it's failed miserably. The world is no darn good. People are no darn good. They're greedy; they're selfish. The Capitalist system has worn out, based on selfishness and greed. But Science can take this to the world in four or five years. Once it is accepted and understood by scientists, the breakthroughs will just be astronomical.

A basic analytical breakthrough. You can fill yourself with this light. In The Dead Sea Scrolls Uncovered, not only did the Qumran community have a metallurgical foundry in the middle of the city, but you also find out that the teacher of righteousness, this thing they were totally preoccupied with, wasn't Moses or Christ. It says the High Priest swallows the teacher of righteousness. The TR is the holy spirit, the TR for those scientists is the light, the zero-point light that isn't measurable. It is, in fact, the light, the god source within us. We know all things. We don't have to read or study. We just know.

When your light body exceeds your material body, you don't have to eat food. You can if you want, but you don't have to. You have perfect telepathy. How much more could you ever be judged than for everyone to know your heart and your mind. Everything about you is known. No more hidden agendas, no more lies, no more deceit. Everything is known. And this is called the opening of the book of light. In Revelations, it says, "Blessed be the man who shall overcome, for he shall be given the hidden manna, the white stone of the purest kind upon which will be written a new name." He will not be the same person. It's encoded in your DNA waiting to be activated.

It says that at 1160 degrees, the white powder of gold fuses to form gold glass. It's a transparent glass, just like window glass. And in Revelations, it says, "The streets of the New Jerusalem will be paved with gold of the purest light, as transparent as glass, and the foundations of New Jerusalem will be made with gold liken unto glass."

This is the gold glass, the very basis of the New Jerusalem. The very basis of raising our self and our consciousness to this higher state. This highest light that will activate all of our DNA will cause us to use all of our brain again, and we will return again to the original state that we were created to be in. Before we fell to the sleeping existence we know now.

These elements are in all of the herbs, the herbal teas, and many of the vegetables you vegetarians are eating. You get them in small amounts. Through work, dedication, years of study and meditation, you can achieve similar results. But it is tough to be a Tibetan Monk. This is called the Keys of the Kingdom. You insert it, and turn it and… It isn't the answer, but the way to the answer. But if you step through that door, that's your decision. Nobody's going to make you take it.

There are many people in this world that don't want this to happen. But this is the New World Order, just not the one George Bush saw. It can be scary. But it is here. Every piece is now known to Science. The philosophical implications are immense.

References:

[1] Sitchin, Zecharia, The 12th Planet, 1976, The Stairway to Heaven, 1980, The Wars of Gods and Men, 1985, The Lost Realms, When Time Began, Genesis Revisited, 1990, Divine Encounters, 1995, Avon Books, New York.

[2] Gardner, Laurence, Bloodline of the Holy Grail, 1996-1997, Barnes & Noble Books, New York; Genesis of the Grail Kings, 1999, Bantam Press, New York; Realm of the Ring Lords, 2000, Media Quest, Ottery St. Mary, England; and most importantly: Lost Secrets of the Sacred Ark, Amazing Revelations of the Incredible Powers of Gold, 2003, HarperCollins, London.

THE ARK IS A BRAIN-WAVE TRANSMITTER AND RECEIVER

It all started over thirty years ago when I wondered what was in the black sand that was washed out of the banks of volcanic mud in Tuxedni Bay. Tuxedni Bay is located between two active volcanoes, Mt. Redoubt to the North and Mt. Illiamna to the South. Several hundred years ago Mt. Redoubt erupted covering a forest and the Crescent River system with a thirty-foot high wall of hot volcanic mud. Over the next several hundred years wind and waves along the beach eroded several hundred square miles of this mud carrying it out to sea leaving a nice, black-sand, beach. I sent a sample of the black sand requesting a spectrographic analysis by a California company. The test didn't show any platinum or gold but it did have some copper.

Several years later I sent a quart-jar, sample to Bremner Technologies of Los Vegas. They use electron guns to melt the sample into a lump of glass. The process tends to impart electrons so that metals lacking electrons like ORME material can then bond into a metal. It is an expensive test because the huge electron gun furnace, which is made in Germany, cost over three million dollars. After the sample cools it is broken up into a fine powder and mixed with a copper-led flux and melted. The precious metals are then poured at different temperatures into small buttons and weighed.

The test of my sample of beach sand revealed 1.5 ounces of iridium per ton. This metal of the platinum group is used in light bulbs and spark plugs. The test also revealed a half-ounce per ton of platinum, three quarters of an ounce of rhodium, a half-ounce of gold, a quarter-ounce of osmium and some copper. At today's metal prices the total of one ton of sand is worth approximately $700. Without the platinum group metals mankind cannot venture into space. These metals are required to make jet turbine blades hard and tough. The skin of rocket motors is coated with osmium to resist heat. Several of the platinum group metals floress giving off intense light and can take surface temperature higher than any other kind of metal. These metals are used in light bulbs and catalytic converters which actually clean up the environment. The reason they floress is because their electrons can be sped up the speeds approaching that of light without burning up. The platinum group metals actually clean up the environment when they are used as catalytic converters.

Over twenty years I read the California Mining Journal and other publications about the platinum group metals, how to extract them from beach sands, how to find it, and how to determine what percentages are viable to mine. The more I learned the more I became fascinated with these metals.

Finally in 1998 I read David Hudson's lectures in Nexus Magazine and learned that half of the periodic table of elements can exist as Orbitally Rearranged Monatomic Elements. The ORME state is where two atoms share their protons and neutrons creating one large nucleus that is eight times larger than an ordinary atom. Due to the fact that the nucleus is so large the elections have to travel much faster than they do around a normal atom to keep it together as a particle. Sometimes they travel faster than the speed of light, which gives the ORME matter inter-dimensional properties.

The whole concept was too fantastic. It was the answer to everything and all my questions. It is a room-temperature superconductor with thousands of amps of electricity flowing through it. This explains why Uzza in the Bible was electrocuted when he reached up to steady the Ark. The Hebrew were transporting the Ark over rough ground in a wagon and it was about to tip over. Uzza reached up to steady it and was killed instantly. The electric shock does not affect those who have been eating the material baked in the priest bread for a long time because they have the same Meissner field or frequency as the ORME material itself.

The refined material becomes lighter than gravity when heated. It is the light of life, curing all disease by correcting the DNA. This discovery makes the platinum group of metals even more valuable. This material is the reason we exist. Our bodies depend on it for cell division, as do plants and all life on the earth. A small portion of it may carry the soul into other dimensions. This explains out of body experiences and a host of phenomena that have confused science for a very long time.

After reading Sir Lawrence Gardener's book it dawned on me how the Ark was used. The one thing I knew that Zechariah Sitchen and Robert Hoagland didn't' know about the Ark is that it contains the monatomic, superconducting gold and platinum powder which the ancient Egyptian priests ate in their especially prepared coarse grained bread. The un-leavened bread takes it through the stomach where it's absorbed by the lower intestines.

No mummies were ever found inside the large pyramids. How could they be built as tombs? There is no evidence that they were ever built by the Egyptians. There is only one Egyptian hieroglyph found inside the Great pyramid and that is a glyph for bread carved over the door of the king's chamber. Bread? There is no proof that the glyph for bread is actually Egyptian. It could be even much older than the Egyptian civilization itself. Why do scholars dismiss this simple glyph? The Egyptians were prolific writers about everything and left long lists and describing every phase of daily life. Given this fact you would think that if they had built the pyramids they would have left some kind of written record about the actual construction. The Ark was placed inside the great pyramid inside the small stone box, which is too small for a coffin. It was carved to hold the Ark. The construction of these buildings is beyond the capability of the modern builders let alone Egyptians who only had copper and stone tools. John Anthony West consulted the largest stone quarry in the world to see if they could build the great pyramid. Their engineers measured and counted the number of stones and came back with a time frame of laying one stone every three minutes for twenty-seven years. They decided that for all practical intents and purposes it could not be built in one 'human' lifetime.

What if a person could live a thousand years? The book of Genesis lists the names of several people who lived over nine hundred years. It is very probable that people using the life-extension, white-powder-of-gold oversaw the construction of the pyramids. According to ancient legend mentioned in chapter one, the pyramid complex was built as a training exercise for Marduk.

One South Carolina engineering firm estimated that it would require using all eighteen of their rock-quarries at peak production and it still would take twenty-seven years to cut the stone blocks for the great pyramid. Then they estimated it would take another twenty-seven years laying one stone block every three minutes to build the great pyramid. In addition the polished limestone on the outside had to be fitted to exact specifications and polished to the tolerances of a contact lens to where light cannot penetrate the cracks. No construction firm existing today can fit 16-ton limestone blocks so perfectly. How did they do it? The people who built the Pyramids knew mathematical concepts and natural laws that go far beyond what we know today. Incorporated into the dimensions of the Grand Gallery is a chronology of all time. The builders obviously could read time seeing down through the ages as if reading a comic book.

Incorporated into the layout of the pyramids and surrounding structures is the latest NASA figure for the speed of light. Men could have gained the wisdom to build the pyramids using incredibly simple and astounding natural technology however it would have taken several thousand if not millions of years of linear evolution. No, they had help from someplace else: Most likely Mars.

As the Bible tells us, Moses was adopted by an Egyptian Princess and grew up to become a high priest. As a Priest he was privy to all the ceremonies and the secrets of how the Egyptians ruled the world for thousands of years. Moses and other adepts could manifest what was called shikana fire on the brass pan or mercy seat built into the top of the Ark. This fire was a corona or electrical-discharge, a holy fire, the voice of God between the outstretched wings of the Angels on top of the lid. This electric discharge was used as a way to test the mental powers of new priests to see how big a discharge they could produce. The Ark itself is essentially a large capacitor made out of layers of wood and metal. The pyramids were built to keep in touch with the other pyramids around the globe by broadcasting messages from one to another.

The material inside the Ark's is inter-dimensional. When one ARK was activated and tuned to the proper frequency shekinah fire would be manifested in all the other pyramids. The frequency or Meissner field of the White-powder-of-gold is the same frequency as brain waves. When someone commands it with their mental powers it causes electrons to flow producing a high voltage. Hugh electrical discharges are produced.

The white powder is a perfect insulator but it responds to a messier field of the same frequency. The brain uses the same material inside its cells to send messages from one part of the brain to another faster than light with magnetic frequency. This material responds to thought. Turning thought waves into a huge electrical corona by amplifying thought waves a million times. No wonder the ARK works like a radio transmitter. It is capable of amplifying thoughts, possibly several billion times. One could send a message to God or anywhere in the Universe instantly because it is broadcasting inter-dimensionally. In some cases the message would be received before it was sent. This is a hard concept for humans to grasp but that is the way space works. Scientists are finding this out by splitting two light beams.

When platinum and gold is in the ORME state two atoms link up to form a one large atom. The two nucleus combine sharing protons and neutrons together in the center. The electrons orbit out at a thousand diameters from the nucleus and link up in opposite rotations, one time-forward electron to every time-reverse electron. At times the speed of these electrons exceeds the speed of light thus giving the material its inter-dimensional properties transcending space and time. That is why the human brain is capable of seeing into the future, trans-locating and a host of other unexplainable phenomena. The soul which has the ability to transcend time and space may be composed of a small amounts of ORME material.

If you want to rule the world simply place the amplified thoughts of a powerful adept using ninety percent of his brainpower inside a giant pyramid fitted together with perfect precision and grounded to bedrock over a large area. It is easy to see how the Egyptians ruled the world for thousands of years. Anything they ever thought came true. They could send their wishes around the globe instantly. Picture a network of these pyramids on all inhabited planets through-out the galaxy and you have a super civilization with god-like powers capable of sharing information rivaling the internet only faster. The messages are inter-dimensional reaching any point in space in one Planck second (10-44 seconds). The builders of the pyramids were able to access the morphogenetic grid of the galaxy.

Recently William Henry and Sir Lawrence Gardner got together and decided that the drawings in the Temple of Man at Abados which depict the ark actually show another device or part or that goes with it. It is shaped like a stele and is mounted on top of the ark in such a way that there can be no electric discharge. Instead, the inter-dimensional electricity is forced into the stele or antenna where it is broadcast in a beam to another place or planet. Could this be the star-gate? If one were to eat enough of the inter-dimensional material and were to stand within a certain distance of the ark could one be transported to another place where there is an identical ark?

King Solomon had a front porch twenty stories high containing a giant stele which he used to transport himself and others to his various business ventures around the world.

PYRAMID DESIGN IS NOT FROM EARTH

While casually looking at the recently released NASA Mars photographs I was astounded by a pattern of evenly spaced triangles on the planets surface. The pattern itself seemed to signify something, possibly a star cluster. What peaked my interest at first was the fact that no two of the triangles were touching each other. If this were some random outcrop of rocks or meteor craters at least one or two of them would be touching or very close to another. Some of them are close but not close enough to suggest that they are a random occurrence. Its as if they were placed within walking distance of each other. The chances of such a thing happening randomly are astronomical. Does this pyramid farm indicate the existence of massive underground accommodations? After turning the page and reading something else the other half of my brain started thinking about architecture in general. I remembered Robert Hogland's analysis of the face on Mars and the D and M five-sided pyramid. I reasoned that if pyramids existed on Mars millions of years ago then the design has to be older than the pyramids on Earth. That would mean that pyramid design is older than mankind's existence on Earth.

Quickly I flipped back to the photograph of the triangles. All of a sudden it jumped out at me that all of the shadows of the triangles were identical. Identical shadows indicate identical triangle structures and if the triangle structures are identical then they have to be artificial pyramids. Why pyramids? Why not build some other kind of structure? Why not build mounds? If you had to build a building or portal entrance for underground habitations that would withstand practically any kind of cataclysmic event such as a flood, winds exceeding 300 miles an hour, or atmospheric depletion you would build a pyramid out of stone that existed near the site. You would make the entrance invisible to keep out intruders and you would build it high enough to be above the floodwaters. If you had no way to glue the joints together to make it airtight you would plaster cement on the inside of the joints. An interesting side note: my son Henry pointed out that the atmosphere of Mars is identical to car exhaust, which is carbon dioxide and carbon monoxide.

A trip to Washington, DC illustrates the fact that much of the architecture in America is copied from the ancient Greeks and Romans who in turn copied them from the Egyptians. So where did the Egyptians come up with their designs? Did they copy them from ancient drawings and texts depicting the buildings of Atlantis, or do the shape of these buildings dictate the use of the materials at hand? With the development of pre-stressed concrete and steel beams, new age domes, suspended roadways and walkways become possible. Are the architectural spires, towers, and flying buttresses of Gothic churches, seemingly inspired by a glimpse of Heaven the only radical departure from the norm? One needs only to examine the inside of the lowly termite nest or the intricate designs of an ant colony to view a similar structure.

Over time the design and construction of thousands of pyramids would be perfected until you had a perfect building. What we have with the Great Pyramid of Giza is a building design that was passed down through the millennia to mankind on Earth. The Mars photographs prove that pyramid design is an extra-terrestrial, artifact. The pyramids are just sitting there waiting for us to figure them out and take the next step, which is obviously to get in touch with our roots. Let us rise up from the tribal warfare over crude oil and get on with the business of becoming a member of the Galactic community, a type one civilization, or as the American Indians call it: 'The Star Nation'.

GREGORY HODOWANEC

Gregory Hodowanec working parallel to Towsend Brown developed a sensitive weighing balance and noticed slight variations in the reference weights he was using. He assumed the problem was in the circuitry he designed. By trial and error he discovered that a capacitor in the right part of the circuit corrected the problem. After much experimentation he found that his weighing system was not at fault. His sensitive balance was actually measuring variations in the gravity field. Sometimes the fluctuations would be at a quite rapid rate. He realized the capacitor was able to somehow pick up these gravity fluctuations converting them to electricity thereby inhibiting his circuit from measuring the fluctuations.

From this information he was able to design a more advanced amplifying circuit. The circuit was connected to a sensing capacitor and the output was connected to a voltage amplifier, which drove a loud speaker. The sounds generated were similar to a Whale's song. Hodowanec stated that his device received monopole gravity waves, which are different to the quadrapole waves described in Einstein's General Theory of Relativity. While Einstein's gravity waves are limited by the speed of light Hodowanec's monopole waves travel to any place in space in one Planck second. Further he discovered that electronic equipment had been receiving these gravity fluctuations for a very long time and had been mistaken for I/F noise.

The universe is filled with I/f noise thought to be background noise left over from the big bang. Forty percent of the white noise on your TV when it is tuned to a channel between stations is I/F noise from space. During his investigations Hodowanec found that Auriga and Perseus in the Milky Way to be the sound of many unusual audio signals. The background noise is modulated by large astronomical bodies, which cast a shadow over the emissions. This means that when the background radiation is de-modulated what you hear is the movement of stars and planets. Stars going supernova, star quakes and even the tectonic movements within nearby planets generate much of the high frequency.

One evening for eight minutes Hodowanec received a series of pulses resembling the Morse code letter S. After determining the origin of these signals, he attempted to make contact using a Morse radio transmitter. To his surprise, he received on his gravity wave detector a reply made up of random Morse Code containing the letters E, I, T, M, A, N, R, K, and S. During the next transmission he received back the same series of letters plus the letters, G and D added. Over a long period of time he was able to carry on a limited conversation with his ETI friends. Arks placed inside pyramids transmit inter-dimensional monopole waves to any place in the Universe in one Planck second. This is the same frequency as Morphogenic grids, which transmit signals from living organisms. Morphogenic grids are faster than the speed of light and non-local. When NASA sends men into space they take a few cells from the astronaut's mouths and put them into a petri dish. Connecting a few instruments to these dishes reveals the emotional stress and the condition of the astronauts. When the astronauts undergo stress the doctors at ground control know it before the bio-med sensors can relay the information back to earth via the telemetry equipment, which uses radio waves traveling the slow speed of light. The kind of information being sent by living organisms from one cell to another is faster than light and the only possible explanation is that cells contain the monatomic platinum group of metals in the high-spin state. The information is traveling hyper-dimensionally. It goes into a timeless dimension, which is not subject to three-dimensional space. Time and distance come out just as if it were a fraction of an inch away. This concept reduces the vastness of space to nothing. After all it (space) came from nothing in the first place. Scientists have discovered that every particle in the Universe is connected with some kind of invisible thread. Following that thread to any specific destination shouldn't be that hard.

I read somewhere that one cubic centimeter of empty space contains all the energy of our sun's output for its entire lifespan. If that is true then we live inside a black hole and don't realize it. Hermetic Law says, "As above so below." Hermes the ancient Greek was probably Thoth the Egyptian and known as Mercury by the Romans because he had some king of space vehicle to travel very fast and deliver messages. He was known as Thoth the Egyptian who wrote the 48 books of Thoth, the emerald Tablet and the Copper Scroll.

I saw a picture of the Copper Scroll on the internet a few years ago but they took it down. It was very shiny copper. They tested the copper and said it was more pure than any copper found on earth.

GIA

The cutting and burning of the rain forest produce morphogenetic waves throughout the Galaxy affecting all living things especially on Earth. I can't imagine what our galactic neighbors are thinking of us when our planet is sends out such hostile signals.

In Gia theory the earth is viewed as a living thing where organisms living and growing on earth influence other living organisms in a positive or negative way depending on many factors. Stressing large populations of living organisms sends out morphic waves, which affect all other living things including mankind. Such stress might be caused by floods, draughts, heat, and hurricanes, which leave large paths of near total destruction.

Thomas Townsend Brown

An article titled Paraseti, ET Contact via Subtle Energies by Gavin Dingley in the January-February issue of Nexus magazine describes experiments by several noted scientists that may have contacted ET civilizations by accident. After a long carrier of working for the military Thomas Townsend Brown returned to investigation petrovoltaics in 1970. The crystal structure of Basaltic and granite rocks somehow converts the pulsing gravity waves to DC potential. The process is well known in electronics and is called:

"rectification." Some rocks have as much as 700 mv across them.

Brown hypothesized the energy is high frequency gravitational radiation which is being constantly emitted from astronomical objects in space. While simple high-k dielectric materials would pick up the radiation and convert it directly into electrical energy, more complex dielectrics such as granite and basaltic rock would convert energy into DC electricity. Not only these rocks are tuned to a portion of the total radiated energy present throughout the Universe. This means that your average lump of basalt is a natural gravity wave AM receiver, tuned into only a few specific "radio stations"!

Brown never analyzed these high-frequency signals to see if any of them were of ETI origin. However in 1953 he filed a patent that described a system for intelligent communication via modulated gravitational radiation. In the patent Brown described how to convert a normal high-power radio transmitter into a gravity-wave transmitter, based upon the principal of electrogravitics. The modification is made to the antenna system, the actual electronics remaining unchanged. A large coil of wire with its base connected to the output of the high-power radio transmitter so that the radio frequency energy is fed. The other end of the coil is connected to a spherical, electrically conducting high-density body. A lead ball works. The spherical body acts like an isotropic capacitor, and so forms a tuned circuit with the coil.

The high voltage and mass of the isotropic capacitor result in electrogravitic action; thus gravitational waves are produced of the same frequency as the end-fed energy from the transmitter which are emitted from the dense, isotropic capacitor. To avoid electromagnetic radiation the whole must be enclosed in a lead lined vault. Producing electromagnetic radiation is illegal because it can interfere with your neighbor's TV reception and airline communication. You don't want airplanes landing at your local airport to crash. Such illegal interference could be dangerous to people, planes, and vehicles.

Interestingly this system is very similar to that employed by Tesla at his Colorado Springs laboratory---the same system that allegedly received signals from an ETI.

PHARAOH

According to the Egyptian Book of the Dead the first thing the Pharaoh did each morning was to go inside the King's chamber of the Great Pyramid and recite the Ten Commandments. These commandments were slightly different from the commandments issued by Moses. Moses most likely borrowed them from the Egyptians. The Pharaoh would say, " I shall not commit adultery." "I shall not covet thy neighbors wife." "I shall not steal. " " Etc. etc." Placing the Ark inside the Pyramid which is essentially a large semiconductor antenna grounded to bedrock over the area the size of eight foot-ball fields it becomes an incredibly powerful brain-wave, transmitter capable of transmitting thoughts faster than the speed of light to the far corners of the galaxy. The pyramids were built as intergalactic communication devices of incredible simplicity. I can think of no more important reason to build such a costly structure than to phone home.

The pyramids were also used as tools to gather incredible power and knowledge from the Cosmos. When the pyramid was being used to control the people everyone within a thousand mile radius heard the Pharaoh's voice in their heads and swore allegiance to the Pharaoh. I believe the age of pyramid technology actually represents a dark age in the history of the human race when all people were slaves to the Pharaoh dynasties for thousands of years. It is a dark secret, something you will never read about in the history books. HARP technology (High Altitude Research Project based in Glenallen, Alaska) is based on Tesla's devices for sending power, electro-magnetically through the ionized layers of the stratosphere. Military scientists convert this energy with a network of satellites to powerful brain-wave frequencies, which they can beam down to Earth commanding troops and people on the ground. We won't have elections anymore. Instead we will just sit and listen to our government masters.

Ancient Egyptian Gods were able to send messages inter-galactically surmounting the obstacle posed by the speed of light. They may even have scalier-wave capability allowing them to utterly destroy cities with intense heat. This is several steps ahead of HARP technology. This ancient regime controlled several star systems for a million years until a major uprising gave birth to the Council.

The Council decided that Free Will should take precedence in all their colonies.

There is a pyramid in China east of Nan King that is larger than the great pyramid in Egypt. There are pyramids in various states of decay all over the world. There are over a thousand pyramids buried in the jungles of Central America. It is quite possible that people once communicated through pyramids. Given the fact that they knew about the power of the anti-gravity, super-conducting, white-powder it is a very real possibility that mankind could send messages around the world and to anywhere in space in one Planck second which is instantaneous compared to light speed. The technology involved in constructing an inter-galactic brainwave transmitter is so amazing that it is inconceivable that any human being could invent such a thing. When you learn as I did that the white-powder gold is inter-dimensional, has life extension and anti-gravity properties then the whole story of how they managed to control the world for thousands of years becomes clear. I feel privileged to be the one who was allowed to re-discover this important piece of history.

Ancient legends proclaim that: "An army that carries the Ark of the Covenant before it is invincible." This can be interpreted in two ways:

1. The moral high ground of those who posses that Ark will insure that they win no matter how high the odds because God and the Ten Commandments are with them.

2. Priests carrying the Ark can transmit commands directly into the minds of the enemy ordering them to retreat and return to their homes. If they did not obey those commands the priests could stop the hearts and bodily functions of the enemy causing them to drop in their tracks.

AN INTER-GALACTIC RECEIVER

Inside the Great Pyramid of Ceops the five 70-ton-granite, slabs mounted directly above the King's chamber are carved in such a way that they amplify incoming signals on an inter-dimensional frequency. The huge stone slabs are smooth on the bottom and left rough on top for a purpose. As incoming signals or waves of energy of inter-dimensional frequency enter pass the stones they vibrate in all directions. When they exit the polished surface on the bottom they become more polarized (vibrating in one direction). After passing through all five of the stones the resulting signal is highly polarized. If there is any kind of modulation of the inter-dimensional signal on the brain-wave frequency the Ark containing the white, monatomic, ORME material becomes agitated producing a replica of the signal in the dark chamber inside the pyramid. Holographic images from other worlds and important messages were received in this way.

THE CLEANSING MYSTERIES

Jesus emphasizes the overwhelming nature and importance assigned to these teachings by the Church of Love in one surviving Gnostic text, the Pistis Sophis. In this text Jesus told his followers: "Stay not your hand until ye find the cleansing mysteries which will cleanse you so as to make ye pure light, that ye may go into the heights and inherit the light of my kingdom."

Jesus is telling us to find the means to transform ourselves into beings of light that we might travel the cosmos. He is saying drop the "I" and see your body as an "it," which is to be transformed like a cocoon changes into a butterfly. This may explain why the church dug up the Cathar's bodies and burned them; they were transformed corpses.

In India, the secrets of the eight great psychic or magic powers which allow the transformation into a body of light and which enable the soul to take flight, are know as the ashta-siddis, which the Hindu and Buddhist masters are believed to have taught Jesus. Among these abilities are levitation, the ability to dive into the solid ground and move about as if it were air or water, and instant materialization of wishes.

These abilities are identical to the presumed contents of Jesus' Book of Love. Mastering them enables the initiate to ascend the evolutionary ladder, become a god-like being, or homo Christos and enter AMOR. Obviously these abilities would be desirable to the Pope's hires warriors, and in fact, spiritual warriors of all ages."

MONTSEGUR: THE CATHARS LAST STAND

As far as I can tell, Montsegur is the most important archaeological site the Nazis ever excavated? (And there are many) Atop its soaring cliff hangs a majestic mountaintop Cathar Seminary fortress strongly reminiscent of Masada or Magiddo.

Beginning here March 16, 1244, was the Cathars last stand. It was at Montsegure that converted papal warriors defended the Cathars. After a siege of several months, more than 200 hundred Cathars were burned at that stake. Some say they ran down the mountainside and jumped into the waiting bonfires in the valley below.

Legends report that a treasure of immense importance was smuggled away from Montsegur in the dead of night by four mountaineers before the rest was massacred. When asked what he believed this treasure was, Dr. Arthur Guirdham, a world famous expert on Catharism, replied that it was most certainly "books, or esoteric manuscripts of some kind." Was the Book of love among these books?

By exterminating the Cathars it was as if the early Church fathers were trying to close the gates of earth before too many souls escaped to heaven. The result of Christianity's campaign, which still persists today, has been the destruction of the secret teaching of Jesus. The faithful should take heart, however. A prophecy says that the Book of Love will be discovered at a preordained time" by a pure heart. It is believed The Book of Love is hidden in a cave in Southern France.

William Henry

"My research developed in my books The Healing Sun Code, The Crystal Halls of Christ's Court and Ark of the Christos, reveals that we now have such a word: stargate or wormhole. The Anunnaki, possibly with E.A. in charge were operating a door to Heaven, in the vineyards at Eschol. Joshua stole its secrets, symbolized by the cluster of grapes.

THE MOTHER SUBSTANCE

E.A., we are told came to earth from Heaven in search of gold, rare metals and radioactive materials, such as uranium or cobalt. The latter were described as the Lower Worlds blue stones that could transmute a person into an EL or Shining one, i.e., a pure one or Aryan.

"As the Sumarian god of smithcraft and alchemy, a priesthood developed around E.A.'s alchemical teachings. They were symbolized by a cluster of three grapes know as the symbol of the fifty essence or quintessence, also called wood, or the Word, the exotic black (hidden, occult) pure mother substance upon which the world was built. The cluster of three grapes was also known as blue apples." "The cluster of grapes later became cryptograms that were extremely important to the Essenes and the Gnostic Grail heretics that they held sacred and secret." "Cathar watermarks showed a serpent hanging on a cross spitting the Word grape symbol. The pictured base-relief from Sion, Switzerland illustrates the medieval belief in the concordance of the theft of the blue apples from the Anunnaki and the Crucifixion of Jesus. Old Testament scenes are grouped together with the Passion of Jesus as if they happened at the same time."

"When put together with depictions of Jesus (E.A.?) As a serpent on the Cross the message is clear: the Gnostic Crucifixion was a stargate or wormhole event..." Grail researcher Andrew Sinclair says: "The original French word for Grail was escuele. Phonetically, escuele, like eschol, is a skill (a skull or skool). This makes sense. Christ was called wisdom. His skill involved a transformation of homo sapien into a pure one, a Cathar." "The blue apples of Eschol (the Languedoc?) Represent powerful physiological knowledge; the means to enter the gate of God."

I discovered later after I put this book on the web that E.A. was Nakosh, the one YHWH punished by turning him into a snake. "Hundreds of thousands of Cathars were murdered because of this ultimate secret. It can happen again."

BIBLE CODE and ISSAC NEWTON

Recently scientists, mathematicians, Rabbis, and Bible scholars have become aware of an equal-distant number code inside the Torah that actually names various people in the future. It reveals the person's death, the name of the person who kills him and in many cases the dates and times they were killed. Sometimes it lists the names of other people who were with him at the time. This revelation is almost too syncretistic for mankind to grasp. The only way this can happen is that a God really did dictated the Torah to Moses and that God can see all time as if reading a book. God is in communication with everything everywhere, or to use a more familiar term he can access the: "Collective unconscious."

It is difficult for man to comprehend even a small portion of the mind of God. To do so makes his thoughts reel in rebellion. A God would have the ability to dictate a document with layer upon layer of hidden codes accurately describing events that occurred thousands of years into the future. Such a document has to be impossible yet that is what we have with the Bible. A scientist in the National Security Administration noticed the code in the first five books of the Bible known as the TORAH and couldn't believe it. Eventually he developed a computer program to decode it.

Every fifty letters of the first two books of the Bible spells Torah. Every fifty letters of the fourth and fifth books of the Bible spell Torah backwards. It's as if they are pointing to the middle book, Leviticus. The middle of the five-tined, candlesticks used in the Synagogues is lit first. This lamp or candle is then used to light up the others. Every eighth letter in the book Leviticus spells YHWH. It's as if the other four books are pointing to the coming of the savior.

Thousands of historical events with surprising details with dates when these events occurred have been decoded in the Bible with equal distant number sequence codes. The death of President Kennedy was predicted and crossed over by the date and the word Dallas. The death of Lady DI with a date, time, and place of her death is encoded in the Bible.

The assignation of the President of Israel is foretold. To have so much detail in the same location about one person is one chance in ten million. Multiply this by the thousands of other predictions involving other people and the consequences become mind-boggling. Other texts like War and Peace have been tested in similar fashion and occasionally one comes up with a persons name by random chance but no other information or details about that person are provided like they are in the Bible.

A long time ago Sir Isaac Newton, the greatest mathematician that ever lived theorized that the entire universe is a code waiting for mankind to decipher it. He also professed that the Bible was a code predicting the future of mankind. Here was one of our greatest mathematicians of the human race who worked out the formulas for a falling body and many other scientific concepts yet he wrote profusely about esoteric things. Bible scholars and Scientists are theorizing that there are other codes besides equal-distant-number-sequence, codes in the Torah. The possibilities are endless. A God with the power to see, predict, and manipulate the future and write it down, obviously planning the coming of the Messiah; such a revelation is mind-boggling. The most remarkable thing about the Bible code is mankind became aware of it at the exact time that the personal computer became available which enables him to see for himself that the TORAH is the divinely inspired word of God. It also came at a time when renewed faith in the word of God is necessary to keep civilization from war. The name, David Hudson the discoverer of the Manna-—the white-powder-of-gold is also encoded in the Bible.

CHAPTER FOURTEEN

GOLD OF THE GODS

For two hundred years we have been eating vegetables grown in the same mineral-depleted soil. We eat meat that is fed grain grown on this same soil. Is it any wonder why cancer, obesity, and chronic diseases are on the increase? We eat all the right foods but our bodies are starving for essential nutrients. We are overweight while and at the same time we are starving. Will we ever be smart enough to figure out a solution to this problem?

The answer is as old as mankind existence on this planet. Take mineral supplements. The Annunaki gods did. In order for the human body cells to replicate they have to rip apart the genes to produce new cells. The process of unzipping the gene code completely without losing telomeres requires the superconducting, monatomic-minerals of platinum and gold which super-conduct electricity. Each time a cell replicates it loses a few of the monocle chains called telomeres at the end of the gene code. When the telomeres are gone that cell dies.

Loss of body cells is what causes aging. If The mitochondria have enough of the right super-conducting, minerals during cell replication it can produce more ATP energy so it doesn't lose telomeres. Theoretically can replicate itself indefinitely. I believe it is possible for a person to live hundreds of years free of disease and without the pain of aging providing it has these minerals.

A God or light being who ingests the inter-dimensional gold and platinum over a period of several hundred years will become a truly inter-dimensional light being to which time and space have little meaning. After a hundred years of eating the white-powder-of-gold your body would not need food and would be capable of existing in any dimension at will. If you eat regular food that decays with time so will your body decay with time? You are what you eat. I have no doubt that the human body can transform over time into a light being having powers to exist in any place or time at will or even trans-locating and even manifest objects at will such as providing food for a group of starving people.

Carried inside the Ark of the Covenant was a white-powder, super-conducting, gold and platinum mineral with inter-dimensional properties. This exotic material is not a metal and cannot bond into a metal because it lacks electrons to form a bond. The electrons of this material have to spin faster than ordinary matter in order to balance the forces within the nucleus because it has the same number of protons and neutrons as two complete atoms compressed into one. This super-sized atom is eight times larger than a normal atom. When in this state the electrons form perfect pairs in opposite rotation frequently exceeds the speed of light. This is what gives MANNA its inter-dimensional properties. This is ORME material or Orbitally Rearranged Monatomic Elements.

Static charge and temperature determine if the high-speed, electrons will exceed the speed of light making the white-powder-of-gold lighter than gravity and creating time anomalies. Individual atoms of this stuff exist in one or more dimensions at the same time depending on the electron spin rate.

It produces a magnetic field with no polarity called a Meissner field. Thousands of amps of electricity course through this material all the times with practically no voltage. It is white because it reflects all light. It is truly the light of life. The ancients baked this material in the bread of the priests and called it manna, the "Bread of the Gods". They extracted the white-powder-of-gold from ore treated with alkaline taken from the alkaline-rich soils of the Nile Valley. The process is simple and can be duplicated in any kitchen providing that you have a source of gold and platinum such as gold ore, seawater or volcanic soil. The word monatomic refers to a single atom or molecule substance that cannot form a solid. The individual particles are too small to be seen with a microscope and are only eight times the size of an atom. It remains a powder of incredible fineness yet one can see and feel it. If heated to 600 degrees it disappears but is still there but in another dimension. This is hard for people to accept. Our thought process is used to dealing with three-dimensional objects and linear time yet such things do not exist in most of the known universe. When we try to conceive of a multi-dimensional substance it is like trying to comprehend the mind of God.

I have been on a quest for the answers to the question; 'what was carried inside of the ARK of the Covenant'? I studied everything on the subject including the Dead Sea Scrolls to the lost books of the Bible. Little by little the answers came to me.

All this information started coming my way over thirty years ago when I walked up the beach on my homestead in Tuxedni Bay and wondered what kind of minerals were in the black sand. After years of wondering what this stuff was I sent a sample to Bremner Technologies of Los Vegas, Nevada? They use an electron beam furnace, which melts the sample with electrons. The tests were encouraging. The beach sand was worth about $750 per ton in precious metals.

I live between two active volcanoes. Volcanic mud that flowed from these mountains produced the flat land where I live. This soil is rich in the platinum group of monatomic minerals and is concentrated when the ocean waves erodes the banks down. Over the centuries the ocean has cut into this mud carrying away millions of tons of soil leaving boulders and the heavy, black-sand, which is rich in iron, gold, rhodium, iridium, and platinum. I eat vegetables grown in this rich soil and drink birch sap from birch trees that grow on it.

Plants use alkaline to metabolize food. I suspected that the process the ancient Egyptians used to extract the white-powder-of-gold form soil was an alkaline process. In 1988 my wife and I drove to Montana to view Yellowstone Park and other sites. While in Bozeman, Montana we happened into a used bookstore. The first book I came upon laying on a table was an 1898 U.S Geological Survey Report for the western states. When I opened it on the counter there in front of me was a picture of Fossil Point in Tuxedni Bay Alaska. I was so astounded that I had to buy the book. I knew it wasn't chance that such a thing could happen. In fact the chances of such a thing happening at random is about a billion to one. The book set me back $75 but from then on I knew I was on a mission given to me by a higher power.

Several months later I was reading my 1898 book in the comfort of my living room when I came upon a description of the mining of beach sands of Gold Beach Oregon. One of the techniques they used to improve the recovery of gold and platinum group of metals was to put the sand into a ball mill and add potash. The steel balls pounding the sand broke open the carbolic shells that form around small platinum fines.

The potash cleans the metals and the kinetic energy of the steel balls hitting the fine grains actually imparts energy by adding electrons. This separates the monatomic atoms into regular atoms of platinum and gold when they are melted.

It all made perfect sense. Energy can never be lost or destroyed. It is transformed into something else. In this case, electrons. Transmutation was taking place even though present-day science says it is impossible. Kirlian Photography works because every living thing including your fingers, plants and leaves have small amounts of the large, monatomic, stealth-minerals with high-speed electrons. These high-speed electrons interact with the contact plate, slowing them down, which changes them into photons.

The negative image on the Shroud of Turin can be explained by Christ's ingestion of the white-powder-of-gold. Intense light radiating from his body chemically reacted with a weak silver solution on the cloth when he translocated outside his tomb. The presence of random inter-dimensional, monatomic, atoms in almost every substance gives every object a presence after it is moved. Auras and signatures left over after an object is moved can be explained by a few of these inter-dimensional stealth atoms still existing in the same place because they are not affected by time.

Much scientific work needs to be done on this anti-gravity material. Time study anomalies, plant growth, time travel, inter-dimensional travel and ten thousand other discoveries are waiting for us. Lay lines, ancient churches and space portals like the Great Pyramid may be located over vast deposits of monatomic minerals buried deep within the earth's crust. Core drilling in such locations may reveal higher concentrations of monatomic than in other places. The earth itself is giving off a modulated gravity wave signature as it rotates constantly sending out an identity signal to some far off place in the Universe. Many people mentioned in the Book of Genesis lived over nine hundred years. When I was a kid I was unable to believe the Bible because I didn't believe that men could live that long. When I was a child I asked my mother about this and she said that they probably counted years differently in those days. There were a lot of things mentioned in the Bible that were hard for me to believe. Now that I know about the white-powder-of-gold and how it works in the body I now believe that certain portions of the Old Testament Bible are true. We know there is a correlation between diet and longevity. Wherever there are large numbers of people who live over the age of one hundred there is a lot of monatomic minerals in the soil. Usually these people live at high altitude and drink milky, glacier-water, which contains rock particles and monatomic atoms.

Why continue to suffer the disease of aging? Now is the time to get in on this tremendous opportunity. It worked for King Solomon allowing him to live over nine hundred years and it made him a billionaire at a time when the world's currency was gold and silver. If it will work for King Solomon it will work for you. Picture yourself helping others to live longer and healthier lives. Picture yourself knowing the meaning of life and being able to understand all truth.

Doctors are treated terminal aids patients with the white-powder of gold and they got well. Many of them went home within a month or two and their relatives never knew they even had a disease. It also works on cancer correcting the entire DNA throughout the body to that of a twenty-year-old.

FATHER CHARLES MOOR

The following are my notes about Father Charles Moor a Catholic priest who was on the Joe Siegel radio talk show November 24, 2000. A four hour transcript of this show can be ordered by calling: 1-800-917-4278 and order tape number: 001124C. The price is $27.50.

"The Catholic Church now acknowledges the existence of extraterrestrials and is terrified of the implications. The Vatican now agrees that Zecharia Sitchin translation of ancient Sumarian texts is correct and is terrified of what the revelation of the Anunnaki Gods presence on earth might do to the Catholic Religion."

According to Sumarian texts the Anunnaki visit earth every 3600 years. The reason they came here last time was to mine gold. Not the heavy, hard, yellow-stuff we think of as gold but the white-powder-gold. They tried to take it out of seawater but the process took too long and the supply wasn't adequate for their needs so they went to South Africa to mine the yellow metal instead.

Moses goldsmith, Bazaleal, made the white-powder-gold or manna by burning it on the mercy seat of the ark. It is being made today by burning gold in an argon filled chamber at temperatures of 5600 degrees. This changes the shape of the atoms from a spherical shape to a conical shape. It then becomes the most nourishing substance on earth where if a person eats it their body converts to the physical age of 26. Can we possibly live as long as the Anunnaki, which live 350,000 years? It feeds the sole and corrects the DNA.

The Anunnaki Gods created the 120-year life span for humans on Earth by restricting access to the bioorganic powder so that the humans wouldn't overpopulate the earth. These revelations may or may not affect the Church because people are slow to comprehend the meaning of the ancient Sumerian translations. It is true that modern humans suddenly appeared in the form of Cro-Magnon man about 300,000 years ago, with no evolutionary evidence.

Archeologists are puzzled by this and cannot offer any reasonable explanation for it. Occam's Razor and Sherlock Holmes concur: "The only remaining answer no matter how implausible must be the right one."

The monotheism (belief in one god) of Bible translations was an attempt to unite mankind in service to a common religion in the hope that it would stop wars thereby allowing civilization to take root. At that time Europe had to crawl out of the slime of the dark ages.

Father Charles Moor went on to say that: "David Hudson, the man who recently re-discovered the white-powder-gold, manna set up a manufacturing plant in Santa Barbara, California. In the first videotape of his lectures that he received form David Hudson. Father Moor said that David Hudson looked like he was in his sixties on the first videotape but in the last videotape he received from him he looks like he is in his thirties. David Hudson has disappeared from public view and cannot be found anywhere.

CODE OF LIFE

Within the Temple of Man in Egypt is encoded mathematical formula for life known as the Fibonacci spiral. This is sacred geometry on the highest order. This mathematical progression governs how plants and animals grow. Examples of it seen in nature are the ram's horn, the spiral pattern of the seed in a sunflower, an artichoke, and various sea shells.

This code of life is a progression of numbers where the next one is the sum of the previous two numbers. Such a progression grows as follows: 0,1,1,2,3,5,8,13,21,34,55,89 144, etc. Even the galaxies themselves are guided by this mathematical principal because they eventually resemble this spiral. This is the genesis code of the universe. If such underlying mathematical principals of nature exist then life itself is concurrent with structure of the universe.

The weird part about this is that modern man cannot see it. It took an ancient Egyptian hiding it in the architecture of a temple to point it out to us.

Where did they get this concept? They most likely got it from the Anunnaki gods who cam from Sirius.

EATING THE MFKZT POWDER

The drawbacks to taking the stuff are you develop telekinetic and telepathic powers. You can read minds and send mental messages. Practically everyone who has taken the white-powder-gold has moved away from cities to a very remote area because they hear and take in all the thoughts and all the suffering of those around them.

The biggest obstacle is how to handle the psychological aspect. The power to bless also holds the power to curse. First we must learn to use the power before we can consider taking the stuff. How do we reach the point to trust ourselves? First you must learn to trust the voices that you hear. The call to become a master is a tremendous decision. Once the power begins to manifest within a person he has to retain conscious thought at all times otherwise he might accidentally cause harm to another. If you are a master healer and wipe out cancer and other diseases saving thousands of lives you still may go wrong. One of the humans you saved could invent a new weapon that kills millions. Life holds no guarantees.

The discovery that the Anunnaki gods lived on Earth does not destroy religion. Religion is about the spiritual bonds that connect us to nature and one another. The very first religion on earth before the Anunnaki came here was animism. An animist is one who discovers that the universe is made out of spirit. Conscience is the stuff that the universe is made of and our conscience is only a small part of it. Shamans have come to this realization thousands of years before the gods came from outer space. In other words, God was busy long before Moses crossed the Red Sea.

TOO MANY PEOPLE ON EARTH

Our government and the news media are constantly telling us that there are too many people on Earth. It is a lie. If we didn't waste our natural resources Earth could sustain a population of 11 billion people. There are currently a little over six billion people on this planet. Now a planet with 11 billion spiritually awakened, people cannot be controlled. Such a scenario would change the harmonic frequency of the whole galaxy. The vibration would be love, and the bad guys don't like that.

In 30 years when the population reaches 10 billion we will be consuming almost everything the planet produces. We are losing a forest area the size of Great Britain annually. Twenty percent of all species of plants and animals are gone forever and in forty years at the present rate of extinction we will have lost fifty percent of all plants and animals on earth. We cannot continue to live beyond our means. The Gods gave us dominion over the animals and even the fish in the sea. When they get back and find out what we have done to the place there will be Hell to pay.

Our leaders are so corrupt that their systems are breaking down. Instead of acting responsibly by releasing control and letting people take more responsibility, the corrupt leaders want to kill off most of the people in an attempt to maintain the status quo. There is a way of speeding up the spiritual awaking and that is what this book is about.

What if a person could live 300,000 years? Think of the long-range goals you could have. You could do the real big projects like genetic manipulation of a species or terra forming. A person could learn every foreign language on earth, or play every musical instrument known to man and become knowledgeable about any subject. Living a long time would tend to lend more meaning to life but it poses a whole different sort of problems. Your car and house would wear out many times before you reached the point where you would no longer need these items.

"What we visualize tends to come true. The Universe is set up this way. This is why it is very important that we visualize a positive future."

CHAPTER FIFTEEN

MORPHOGENETIC FIELDS AND INTELLIGENT PLASMA

Rupert Sheldrake wrote in his book: DOGS THAT KNOW WHEN THEIR OWNERS ARE COMING HOME, AND OTHER UNEXPLAINED POWERS

Published by Random House: proposed the existence of subtle, hyperspatial: "morphogenetic fields" which guide the formation of matter or living systems. The fields are further strengthened by the physical manifestation they help to form. This makes it easier to repeat the physical form.

For example in chemistry it is often difficult to grow a new crystal for the first time; but after one laboratory succeeds, it is easier for others to accomplish this, even at remote locations. A process that previously failed begins to succeed after the first success. Elementary particles react the same way. Once the first one is created in the accelerator it becomes easier to generate more anyplace in the world.

Sheldrake's theory states that the crystal or particle has a morphogenetic field which becomes "locked in" at the first physical manifestation. This field then guides future growth and creation. The field is non-local and hyperspatial in nature and can be likened to an "ethric" or "spiritual" form.

Morphogenetic fields guide the mitochondria

The mitochondria that live within the protoplasm of a single cell as autonomous beings—yet they participate in a collective fashion to provide the cell with energy. The morphogenetic fields of a single cell guide the collective behavior of its components. The same goes for the cilia of a protozoan.

The morphogenetic fields link the separate individuals of specie. When there are only a few termites their pattern of moving pellets is random and meaningless. When more termites are added to the group a threshold phenomenon occurs. Their behavior radically changes and they begin to cooperatively create majestic, multi-arch structures for their nest.

Another example is squid migration. When a few squid are gathered, there is no awareness of what direction to swim; but when a sufficient number are present, a new group intelligence arises, and the collective acts as a single organism making a direct, purposeful migration across the ocean.

In the human brain memory is stored redundantly on many neurons. The ability, clarity and quickness of recall are related to the large number of neurons. Memory is not localized in the brain, but redundantly distributed.

Scientists studying the intelligence of monkeys on Pacific islands near Japan noted that they refused to eat sweet potatoes because of the sand on them. A scientist taught one of the monkeys to wash a sweet potato and it began to consistently wash and eat them. Soon, by imitation the other monkeys learned to wash the sweet potatoes .Soon all the monkeys on all the islands were washing their sweet potatoes.

The message became encoded in the collective group mind of the species and this mind (or morphogenetic field) was non-local within space-time. The scientists were startled to discover that all the monkeys on all the pacific islands now knew how to wash the sand off their sweet potatoes.

V. Woolf extends the concept of a non-local collective group mind to human beings. Woolf describes the processas "holodynamic" psychology. All minds are in a constant dynamic state of growth, yet all are a part of the group collective or holistic universal mind. Each human mind, in turn, is comprised of many more primitive minds called "holodigms"—each with its own ego that experiences separation from the other holodigms. What we experience as our ego is simply the holodigm that is currently active or conscious. The word "holodigm" means whole (holo), form (digm). It implies that each primitive ego state is a form that arises from the holistic, universal mind and contains the potential for reconnecting its awareness back to the universal mind.

The process for establishing this reconnection or awaking is called psychomaturation.

This process not only yields a happier, more fulfilled life, but also unlocks the psychic potential of the individual. When many minds are focused together in the psychomaturation process, the awareness accelerates no only for those experienced, but for those who are just beginning the process. The more bonded minds participating the more rapid the growth.

Thus, as more people awaken to their full potential selves (the spiritual-self which transcends the physical body), the easier it will be for others to awake. When a sufficient number come into experiential awareness of the universal mind, a threshold will be breached in the morphogenetic field of mankind, and all minds will spontaneously become universally aware. At this point, all individuals will realize and directly experience that we are a single superconscious entity.

The morphogenetic grid actually connects all life in the galaxy. Quantum mechanics shows that every elementary particle and, therefore, all matter is formed in the zero-point energy, and Sheldrake suggests that existence of subtle, hyperspacial morphogenetic fields which guide the hierarchical organization of matter and living systems. It is hoped that this brief chapter sheds light on the growing awareness of the holistic paradigm and motivates study to usher in a unifying transition for humanity.

It is my feeling the platinum group in the ORME state is responsible for the transmission of information faster than the speed of light. I hope NASA kept accurate records of the time difference between the skin cells of astronauts showing stress in a petri dish on earth and the biomed sensors relaying data back to earth via the telemetry equipment.

DR. GARY AND LINDA SCHWARTZ

Dr. Gary and Linda Schwartz recently wrote a scientific paper on Systemic Memory. The theory is that everything is connected in the universe with a high-frequency tone of around 100,000 Hz which sends messages back and forth each time adding to the message thereby creating an identity for each object and an awareness by other objects of the first objects existence. Boundary layers between skin, and individual cells energy in space all are preserved in objects and space. Once a feedback loop is started it creates self-awareness loops, which are preserved forever. The entire universe is set up this way.

One way of demonstrating this interactive feedback phoneme is to sit between two mirrors placed so you can see your yourself reflected in the second mirror reflected back to you from the first mirror. Your reflection is bounced off the first mirror which in turn is bounced off the second mirror which in turn is bounced off the first mirror but since the image is further away each time the image keeps getting smaller and smaller but it theoretically goes on forever down to nothing. The first mirror is aware that the second mirror reflected the image back of the first mirror and it is aware the information was reinforced by another reflection adds infinitum.

Science is on the verge of discovering the meaning of the universe.

CLIVE BAXTER

One-day FBI agent and lie detector expert, Clive Baxter attached his lie detector to his rotodendrum in his office. He discovered that the plant responded to what he said and other people in the room. Not only that it responded to what he thought.

All life on earth both plant and animal is related and perform similar cellular functions and repair using monatomic minerals. The platinum group of six metals and gold in particular conduct electricity at optimum levels thus they are required more than the rest.

Michael Heizer who I mention elsewhere in this book makes note of the following phrase from Genesis: "Let us make man in our image" as having two plural words. 'Us' and 'our' are both plural. This means that several gods working together created man. Dr. Heizer makes note of the following translations: Kheim is the singular word for God and translates as ALWAYSE WAS. Eloheim is plural for Gods as Malakheim is plural for Angels.

Free Energy

Why are we burning up all the oil, gas, and coal dumping millions of tons of carbon into the atmosphere when we should be using it for clothing and housing? Oil products can be re-cycled and used forever. For some reason our government doesn't want us to do that. They want to hold the world in a status quo state. Common sense tell us that if we burn all the oil and half the coal the atmosphere will revert back to what it was before the plants and plankton converted the carbon dioxide into oxygen billions of years ago.

We will be dead from toxins long before the air quality becomes un-breathable. A change is warranted.

We have the technology to unlock human conscience; expand the use of the brain from ten percent to eighty percent or higher. Why must we all remain stupid? If a commercial fisherman like myself can build a processing plant to manufacture the white-powder-gold and mind expanding foods why not the world?

We live in an age when our basic concepts of reality and beliefs are constantly being challenged. Nothing is ever what it seems anymore. Our government operates a separate shadow government perusing secret operations that are not in the best interest of its citizens. What if the citizens of the World suddenly started using ninety percent of their brains and became spiritually awakened? Try to imagine how useless governments would be if there were six billion enlightened people. They want to keep us stupid so that they can control us.

THE ULTIMATE FREE ENERGY DEVICE

Place a small amount of ORME material into a glass tube with an electrical wire inserted in each end. Energize the material with brain wave frequency to make the electrons flow. The device could be made very small or large depending on the amount of power needed. There should be no waste heat. If there is then the unit will have to be made larger and a coolant circulated over it. (Extreme caution must be used with this device because we are dealing with inter-dimensional powers we cannot yet begin to comprehend.)

Several ancient Egyptian hieroglyphs depict a device about a foot high connected to something that looks like a capacitor with extremely heavy electrical cables. The electric cables go from the capacitor to a large glass-bulb about four feet long and a foot in diameter. This glass bulb has a snake in it but it could represent an electric eel because there are electric eels in the Nile. I think it is a large electric light powered by the high current of the white-powder-gold, which is producing shikana fire. Just how they modulated it is still a mystery.

Pictures of ancient Egyptian electrical devices are incomprehensible to modern scientists because a device smaller that a cubic foot producing large amounts of electricity seems impossible. They had to be using some kind of advance technology we are not aware of.

Archeologists examined the walls of every tomb and pyramid in Egypt have not found any trace of carbon from the torches that would have been needed to construct the buildings. The only carbon build-up is from torches that have been used by 'modern' explorers. How did the Egyptians carve the hieroglyphics in the tombs without torches? Surely they must have had some kind of light because they couldn't have done it in the dark. It has been suggested that they used a series of mirrors to focus sunlight into the long dark passages however; some of the passages are so far back that this system of lighting would have been impractical.

According to some Egyptologist they did have electric lights and perhaps an infinite supply of free energy. Dr. Bruce Goldis has an article in The January 2000 Coast-to-Coast News Letter describes ancient Egyptian television. The number for subscription information is: 888-725-5505.

VIKTOR SCHAUBERGER

Viktor wrote extensively about water and designed implosion technology motors. Water is a polar molecule meaning one side is more electrostatically positive charged the other being more negative. A liquid, by common definition, consists of molecules in constant motion.

Combining just these two properties: i.e. constant motion and the polarity will bring about alternate attraction and repulsion. We have an extremely complex medium, much more complex than any mainstream textbook has ever described.

So what does this mean? We are faced with a global water crisis in the near future due to lack of understanding of this complex and powerful substance. An understanding of this can help us maintain our health. Viktor Schauberger studied water and the earth's natural process intently early in the twentieth century and predicted most of the problems witch we now face with our natural resources, especially water. He was very vocal against what he considered to be the devastation of the once vital great rivers of the world by conventional river engineers and hydrologists who lacked his understanding of water's natural process. Viktor wrote on the subject in an essay he called The Nature of Water (1932) where he explained that attempting to regulate water by manipulation the banks of the river was "fighting cause with effect" as the banks by definition must be created by the flow of the water and not vice-versa. He explained why these regulatory processes in every case had devastating effects on the natural vitality of the water causing increases in temperature causing increased tractive forces in "regulated" waterways causing more gravel movement and an overall more murky character to the water. His accounts were wholly accurate yet almost entirely unheeded.

Schauberger found that water is densest at a temperature of +4 degrees Centigrade, explaining that this is an anomalous characteristic in that all other fluids become consistently denser with cooling without exception, except water. Schauber taught us that water molecules in constant motion can vary significantly in momentum and frequency allowing water to be more resistant to flow or more fluid. This results in considerably different physical qualities of water, which vary depending on the conditions that the water has experienced. He call this characteristic of water "memory" and explained that the water molecules, while spinning, have an associated frequency (cycles per second) and will absorb these frequencies from their surroundings.

He characterized water as a "living" substance, considering it the blood of living Mother Earth.

He found significant influences on this frequency "memory" to include pressure, temperature, chemical residues, (even if biologically activated appear to be absent). When chemicals and

Microorganisms weren't actually present in the water their residue still retained a structure or frequency characteristics inherent to the chemical or microbe. Water with the residual cluster structure seemed to mimic the associated biological reactions of the chemicals and induce growth of microbes when consumed by both plants and animals. That is sick water.

Nature maintains high frequency vitality in what Schauberger considered "ripe" spring water by creating meandering stream channels and always flowing in circular paths. Municipal water, on the other hand, is recycled many times through hundreds of miles of strait metal pipes under high pressures, which are never present in nature. These conditions literally dampen water's natural high frequency memory or momentum. Slower spin on water molecules, or clusters. This means the influence of our municipal water system makes water that is literally more viscous (measured scientifically as a higher surface tension) rendering it biologically less absorbable to cells (biological fluids have a lower surface tension). Viktor Shauberger considered well water and distilled water (reverse osmosis water having similar properties) as "juvenile" water that is depleted in both mineral content and high frequency "memory" and saw problems associated with consumption of this type of water as well.

Viktor's discoveries were so paradigm shattering as to literally challenge the entire foundation of modern physical science-not just one or possibly to disciplines, but all of it. His own humility led him to write in the Journal Implosion the following statement.

"They call me deranged. The hope is that they are right. It is of no greater or lesser import for yet another fool to wander this earth. But if I am right and Science is wrong, then may the Lord God have mercy on Mankind. (Page, 28 Journal Implosion)

Viktor constructed an 11 kW home power generator that ran on water. "Electric power generated in my home power generator would make nuclear fission not uneconomical, but ridiculous." (Pg. 12 Journal Implosion)

The "purely categorizing compartmentally" pseudo- education, already in existence then (and much further developed today) stultifies creativity and destroys the "ability to see things as they really are and thereby their connection with nature."

Rivers and streams cleanse themselves, revitalizing themselves by the meandering course they seek out. We are always straitening a river's course or damming it. In Mysteries of Asia video it shows knobs or what scientists call lingums (phallic symbols) carved in the bedrock for many miles and under many rivers and streams. I can't get over how scientists labeled this non-toxic water- purifying, device a phallic symbol. Not only is this an insult to the intelligence of the Asian people who carved them it is the height of ignorance. It seems they wouldn't recognize a high technology device if they saw one. Maybe we should send our archeologists to engineering school. I knew right away that they weren't phallic symbols and had something to do with water purification.

Viktor designed and had constructed two flying saucers. Other groups built several after these prototypes. One of these later models was launched in 1945 and in three minutes, rose to over 58,000 feet, flying at a forward speed of 1320 mph. No electricity was used, except to start its rotation.

Viktor advocated the use of wood and copper coated implements to till the soil and serve food. He devised plows that neither generate undesirable currents in the soil nor facilitated the development of pathogenic bacteria as steel and iron could. Viktor described the mid-western landscape of the United States as a "dying land," dry, hot and treeless as it was, irrigated by "firewater" from above ground tanks. Viktor felt the human consciousness was adversely affected by food grown under non-optimal conditions as these and utilizing artificial fertilizers, which destroys the soil.

Viktor was convinced that the use of cast iron and steel pipes, which had replaced wooden and stone conduits, degraded water quality, contributing to illness, specifically cancer. Water flowing through a cylinder destroys the inherent structure of wholesome water, from mechanical processes associated with the chaotic movement of its molecules and the resultant friction against the pipe's walls, which is very high.

Viktor argued that the natural world arose through the interaction of subtle energies deriving from a fourth and fifty dimensions. These energies were potentiating, of very high frequency, immensely powerful, not sensed by us and dynamic, in an easily, destabilized equilibrium. Many of our scientific "laws" were invalid because we did not take will, spirit and these higher dimensional energies into account.

The law of the conservation of energy is incomplete, motion and energy accounted for all that was, exceeded greatly our three dimensional models. We don't understand magnetism and electricity not the nature of water, the atmosphere, the sun and temperature-all of which h considered at length. The Universe is set up to work by intent: designed by a living, loving, God. The sooner we understand this, the better.

CHAPTER SIXTEEN

TRANSMUTATION HOW TO MAKE GOLD

Several years ago I discovered writings on the alchemy of transmuting led, silver and mercury into gold in Mysteries of Mind Space and Time volume 1. This book tells the story of a 17th century scribe named Flamel.

One day Flamel found a very large, ancient brassbound, book titled, 'The Book of Abraham the Jew'. The cover of this book carried a warning that only those who were learned and had read and studied the TORAH should continue reading the book. The warning went on to list seven kinds of damnation to anyone not qualified.

Since Flamel was a scribe and had read the TORAH he didn't hesitate to open it. Twenty-one years Flamel tried without success to find someone who could translate this book and explain the pictures to him. At last his wife;

Permanelle suggested he should travel to Spain to seek out some learned Jew who could shed light on the matter. He set off at once and decided he should make the famous pilgrimage to the shrine of St. James at Compostela first. From there he made his way to Leon where by chance he met Master Canches, a learned Jewish physician. When he saw the pictures that Flamel had copied from, the book he was overcome with rapture and joy. He recognized them, as parts of a book that he had long believed to be lost.

The book was written to help Jews pay the Roman tax with gold. The two of them set off at once for France but at Orleans the physician died. Flamel, seeing him buried returned alone to Paris. Three months later Flamel made his first transmutation into gold. During the next several years Flamel and his wife endowed several hospitals and churches. Every so often during the next two hundred years Flamel and his wife were seen at opera's and on trains traveling the Orient.

The pictures in the Book Of Abraham the Jew depict men gathering something from the soil in the garden. They were copied so many times that it is difficult to know what they are doing. The other engravings are esoteric in meaning and impossible for anybody to translate unless you have studied the Qabalah for years. What were they gathering in the garden?

I always suspected the extraction process was an alkaline process. After I read about David Hudson' s process it confirmed my suspicions. What they were gathering from the top of the soil was alkali sand. Thus began my search for the Holy Grail.

GERMAN GOLD

In 1924 Hitler was sent to prison for organizing and armed uprising. One of his fellow conspirators, a General Erich Lundendorff, was acquitted and the following years stood for election as President of the German Republic. He was soundly defeated by the national hero Hindenburg; he turned his attention to raising funds for the Nazi Party. There were rumors that Tausend had succeeded in making gold by transmutation. Lundendorff got together a group of industrialists and businessmen to investigate the matter. One of the group was a merchant and he bought the necessary materials, some iron oxide and quartz. sand.

Tausend melted them together in a crucible then took it to his hotel room overnight so that nobody would tamper with it. In the morning he heated it again in his electric furnace in his visitor's presence and added a small pinch of white powder. When the crucible cooled the lump of material was broken open and a nugget weighing a quarter- ounce was inside.

Ludendorff was ecstatic and formed a company, which he named Company 164, a number, twice that of the atomic number of gold. Tausend was to receive 5 percent of the profits and Lundendorff seventy-five percent. The money poured in and within a year General Lundendorff put 400,000 marks into the Nazi Party funds. Then in December he resigned leaving Tausend responsible for all the debts. Tausend continued to raise money and on June 16, 1928 made 25 ounces of gold in a single day. On the strength of this he was able to issue share certificates with a value of 22 pounds of gold each.

After a year went by and no gold was produced Tausend was arrested for fraud. There was a sensational trial and on February 5, 1931. He was sentenced to four years in jail. While awaiting trial he was successful in producing gold while under supervision by the Munich Mint. However, this evidence was contested in court and he went to jail.

Dunikovski, a polish engineer discovered a new way to make gold. He took quartz sand and spread it over copper plates. This he melted with 110,000 volts of electricity and irradiated it with what he called Z rays which transmuted quartz into gold. Investors poured two million franks into his process. After a few months he was fond guilty of fraud and sent to prison. His lawyer was successful in getting him released after two years. He traveled to San Remo in Italy where he continued his work. Soon there were roomers that he was supporting himself by selling small lumps of gold.

Dunikovski's lawyer sent Albert Bonn, a chemist to observe the process. He found that the quarts used contained small amounts of gold but the process used produced a hundred times as much. Due to power limitations the amount of quartz used was small so only small amounts of gold were produced.

In October 1936 Dunikovski demonstrated his process to an invited group of scientists. He proposed that all minerals contained embryonic atoms that were undergoing a transformation in nature that took many thousands of years to complete. He claimed that his process merely accelerated the natural growth of embryonic gold in Quartz.

As a gold miner and prospector with many years experience I arrived at the conclusion that gold nuggets are electrically grown one atom at a time in some streams. This is not as crazy as it sounds because we know that electroplating works and it is done in a very short time. Flowing water carries considerable amount of electricity and it also contains minerals, which are an electrolyte of sorts.

Some springs and streams have a high gold content because the water issues forth from underground gold deposits. The gold content of a stream off water can be measured. The water is filtered through activated charcoal treated to collect gold. After a certain amount of the water is pumped through the charcoal filter it is removed, dried, mixed with a copper-lead, flux and melted. The gold is removed by a cupol process and weighed. If you know how much gold you have and how much water was filtered through the charcoal you can calculate how much the water is worth per gallon. Ten cents a gallon is worth processing in this manner.

Dr. Patric Flanagan, during his search for the properties in Hunza water put water into a blender spinning at about 1000 revolutions per minute. With a wire placed in the middle of the vortex and another wire placed alongside he was able to measure a charge of 10,000 volts. In a stream with swiftly flowing water charged with gold atoms small gold nuggets can grow into much bigger ones in a few hundred years.

DUNIKOVSKI

I have to agree that Dunikovski's process could work. After the war broke out little was heard of Dunikovski. There were some rumors that he established a factory on the Swiss-French border. When the Germans occupied that area there were rumors that they manufactured gold to bolster their economy.

Dr. Archibald Cockren, a respected osteopath, practiced gold therapy. Using antimony, quartz, and iron, he was able to produce gold. He used a liquid called 'oil of gold' on his patients. This same remedy was used many years after his death from a bomb blast.

Ordinary quartz sand is the key. The primary element of quartz is silicon, which is a metalloid, belonging to the group of elements, carbon, germanium, tin, and lead. Quartz is an almost pure compound of silicon and oxygen called silicon dioxide. We now have the 'Rutherford equation':

Si(14) + Si(14) +Sb (51) = Au(79). Antimony Sb is the other element although I believe that lead or iron; atomic number 26 can be used as long as there are monatomics present. Two iron atoms or two times the atomic number of 26 equals 52. Add that to two atoms of silicon, which is 28 and the result, is 80. All you have to do is loose one proton in the process and you have gold.

A good source of antimony is battery lead. Which has antimony content of 4 to 11 percent. It is added to increase the corrosive resistance of grids and terminals. It also increases the hardness of lead.

The bearings in your car are a mixture of lead, tin, and 4 to 18 percent antimony. If a person needed a free source of antimony either of these discarded items are available. Antimony is added to type metal to make it harder. You might be able to get some from your local newspaper. Used type set contains 13 to 30 percent antimony. Don't tell them that you want to make gold.

There is a young man in Talketna, Alaska who will take your gold mining concentrates, put them in a crucible, and melt them with a flux. At a crucial time he adds a small pinch of white powder. A small cloud forms over the melting pot and miniature lightning bolts arc over and through the molten material. When the process is finished you get back six times more gold than if you had sent your concentrates to a professional gold refiner. This man said that the secret of the white-powder-of-gold or philosopher's stone came to him one day when he was reading the Bible.

In the monatomic state (ORME state) two atoms combine, actually sharing their protons and neutrons creating a nucleus that is eight times larger than normal. This nucleus is eight fermies in diameter instead of the usual one fermi. Due to the fact that there are double the number of protons and neutrons the tremendous forces existing at the center of the atom force the protons and neutrons further apart. If you try to hold two magnets of the same polarity close together you know what I am talking about. The nucleus also spins faster than a normal atom.

We have all played with magnets and know that the further you hold two magnets apart the more it weakens the force. The same thing happens inside the atom making the monatomic material easier to tear apart or in this case transmute into other elements. This is why low-power transmutation of elements is possible. The ancients really did know to transmute lead into silver and silver into gold. All they had to do was find the right material or philosopher stone with the right vibrator frequency to turn the lead into the monatomic state. The next step was to use it with other elements rearranging the protons and neutrons to make gold.

Electrons normally travel around the nucleus in pairs: a 'spin forward' electron and a 'spin-reverse' electron. But when these come under the influence of a high-spin nucleus, all the spin-forward electrons become correlated with all the spin-reverse electrons. When perfectly correlated, the electrons turn to pure white light. It is quite impossible for the individual atoms in the high-spin substance to link together. Hence they cannot reform as metal, and the whole remains a glowing white powder. The truly unusual thing about this white-powder-gold is that, when heat is applied, its weight will rise and fall several times above its optimum weight, and down to less that absolutely nothing. Moreover its optimum weight at room temperature is actually fifty-six percent of the metal weight from which it was transmuted.

This white powder is the powder of the highward firestone capable of raising human consciousness, but it is also a monatomic superconductor with no gravitational attraction. This is truly the meaning of life and it is the one substance we owe our material existence to. The genes in our bodies cannot replicate without the super-conducting white-powder-of-gold. Without it our body would surely die.

THE FOLLOWING IS FROM DAVID HUDSON'S 1995 INTERVIEW

"...Question - I mentioned in the newsletter a couple of things that I found very fascinating when you and I were having our conversation. One was about the fact that there was a Meissner field that enabled the priest to approach the Ark of the Covenant without being killed. And I'd like if you would if you could tell us a little about that. And also about the air lift of the Black Jesus and what you think that was about?

"Well, ah, I felt I talked enough about science, so I didn't want to go anymore in that way, I wanted to get to the philosophy because that's what everyone wanted to hear anyway. You got to talk about the science so that the reality of it all is there. But, in fact, when you understand that superconductivity is not electricity. Superconductivity is like a world of its own. A material that is a superconductor literally, a superconductor contains one vibrational frequency within the superconductor--one vibrational frequency much like a laser. This light flows perpetually within the system. That nowhere in the system is there any voltage. So you can't hook up a wire here and a wire here to the superconductor and get current to flow in and out of the superconductor, because to get current to flow off the wire, you've got to have a voltage to get current on the wire, you got to have a little voltage and yet by definition a superconductor won't allow any voltage. So the material's a perfect insulator not a superconductor.

But if you resonance frequency tunes the wire so that the electrons vibrate at the same frequency as the superconductor, then the electrons will flow on as light, as electron pairs. They will pair up and flow on, because they're seeking the path of least resistance, which is the superconductor. Okay?"

DAVID HUDSON'S LECTURE

"We found that the definition of a superconductor is that it does not allow any voltage potential or any magnetic field to exist inside the sample. So, by definition, a superconductor will not allow any voltage to exist inside the sample. To get electricity off a wire requires a voltage, and to get electricity back on the wire requires a voltage. So now I know what your question is, "So what good is this stuff?" Or you can't get energy into it and you can't get energy back out of it, what the heck good is it? Well, what you come to find out is that in the superconductor there is a single frequency of light, just like a laser that is flowing perpetually inside the superconductor and when it flows inside the superconductor it produces around it what is called a Meissner field, which is unique to superconductors.

A Meissner field excludes all external magnetic fields from the sample. What color must it be? It has to be white. Anything that excludes all light from the sample has to be white. Anything that absorbs all light has to be black. If it reflects all light it has to be white. Now, I am talking about a pure single element superconductor. It has to be white when it is super-conducting.

What you have to do is take a radio frequency transmitter and resonant-frequency tune the superconductor to match the frequency of the wire. So now the wire is in occultation with its electron waves, exactly the same as the superconductor. At that point the electrons pair can go on the superconductor with no push at all, because electrons are continually moving on the wire, seeking the path of least resistance, when you have them in perfect synchronization with the superconductor they go on with no push at all as pairs.

This takes a little explaining because on spin-on half electron plus one spin-on half electron are two particles. Yet when these two particles become perfectly paired as mirror images of each other, they lose all particle aspects and they become nothing but pure light. This doesn't make any sense either, does it? But that's the way it is. Spin-one half plus sin-one half gives you spin one, which is now pure light. Trust me, it is so. So they can't go on as individual electrons; they go on as light.

The crazy thing about electrons is that one electron can exist in one space-time, and if it moves to another space-time it gives off light or absorbs light. It's moving from one space-time to another. Now we have light, which is two electrons. Light doesn't exist in any space-time. You can put 50 billion lights all in the same space-time and it is okay. Now we don't have a conductor. If you put electricity on the wire, you've got to take the electricity off or it won't flow. You've got to ground it right? With a superconductor that's not so. It can go on, and go on, and go on; it doesn't have to come off. If you want to take it off, you have to put a wire nest to it and you resonant-frequency tune the wire to match the superconductor. And when it's in perfect harmony you apply a voltage and, 'poof', off goes the energy." There's this fellow by the name of Hal Puthoff, who worked over in the Bay Area in California doing distant viewing experimentation and in now working down in Austin Texas at the Institute for Advanced Studies. He actually developed the mathematics for Sakharov's theory of gravity and published it in one of the top science journals.

In the mathematics he shows that when matter begins to interact in two dimensions, as opposed to interacting in three dimensions (by definition, a super-conductor is a resonance-coupled quantum oscillator resonating in two dimensions, not three dimensions), it should theoretically lose four-ninths of its gravitational weight. Did you know that five-ninths is 56 percent, exactly?

I decided, "I've got to go down and see Hal Puthoff. I've got to take all my data and go down and see Hal Puthoff." So I did, and I said to him, "Hal, we have the experimental confirmation that, in fact, your mathematics are absolutely correct. In addition, Sakharov's theory of gravity is absolutely correct, because the material only weighs 56 percent when it goes to the super-conducting state. Hal Puthoff said, Dave, you do realize that gravity is what determines space-time? When this material only weighs 56 per cent of its true mass, do you realize that this material is actually bending space-time? Now if you think about this, it seems correct.

He said: "Dave, what we really need is a material that totally bends space-time, a material that has no gravitational attraction at all. less than zero."

I'd been reading papers on vacuum energy. Do you know that there is an overlap between the thermal spectrum and the zero-point spectrum? The two of them overlap. So if you heat something, it should interact with the zero-point energy. Well, because this material is resonating in two dimensions, when you heat it, it literally loses all gravitational attraction. You know what Hal Puthoff said to me?

He said: "Dave, at that point you shouldn't be able to see the material."

I said: "Correct. You can look in the pan through the quartz tube and there is nothing in the pan. But the pan isn't weighing what it would weigh if the stuff wasn't in it." Now I had mistakenly assumed that the material was just resonating at a frequency we didn't perceive.

He said: "Dave, theoretically it should be withdrawing from there three dimensions, It should net even be in these three dimensions."

I said: "Wow!"

He said, "Dave, you have to devise an experiment where you can do this: while it is not there, pass and arm through the sample pan. If it is there and resonating at a frequency that you don't perceive, you knock it out of the pan-—because when you cool it down and it begins to reappear, it always appears, in the same shape and place it was in before it left. That's proof that it left these three dimensions." And he said, "Dave, if you do that you will never ever want for money."

"In 1988 I not only filed a patent on ORME's (Orbitally Rearranged Monatomic Elements), I filed patents on S-ORMES, the resonant coupled quantum oscillation system of many atoms of these ORMEs. I have 22 patents."

CHAPTER SEVENTEEN

ANTIGRAVITY

Due to its high speed electrons ORME material is anti-gravity. One cannot help speculate that the legends of flying carpets might be real considering that a quantity of this material sewn between two carpets and charged with high-frequency electricity, via a simple hand-crank, device would lift heavy objects.

How they steered the thing is beyond my powers of speculation but anyone smart enough to figure out how to build a flying carpet in the first place can figure a way to steer it. I can see wrapping it around stone pyramid blocks to make them light enough to be carried by hand and then towing them into place with ropes. A crew with a hundred of these carpets could lay down a thousand five-ton, stone-blocks in one day. The three million blocks of stone in the great pyramid would take 3,000 days or 8.21 years.

To this day scientists cannot fathom how they built the pyramids. Engineers from the largest stone quarry in the world estimated that it would take all thirty of their quarries 27 years to produce the blocks in the Great Pyramid. They weren't sure that they could cut and polish these blocks to the same exact specifications as in the pyramid, let alone hoist them up and place them in position at the rate of one block every three minutes for twenty-seven years. The blocks of stone fit so closely together that one cannot insert a credit card between them.

The measurements of the pyramids are astoundingly accurate based on earth measurements. The people who built them even knew about the earth's equatorial bulge and included it in their calculations. The units of measurement used in ancient Egypt matches perfectly the measurements used in ancient India and many other places in the world.

There are so many details that people have in common all over the world that ancient Egyptians must have had communicated with them. The pyramid measurements are so accurate in fact that it should be obvious to all the people studying them that the people who built them were more

technologically advanced than our present state of development. Only recently numbers corresponding to the Planck second have been found encoded in the Great

Pyramid of Giza
ZECHARIA SITCHIN

This famous author who translated the Ancient Sumarian texts says the pyramids were built as beacons for space travelers. The seventy-ton, granite, spirit-stones above the King's Chamber are still emitting a strange electric signal. The pyramid is sending modulated gravity waves strait up in a cone configuration above the apex. As the earth rotates this signal is beamed out in a circle every 24 hours. Far out in the cosmos space travelers would receive one brief signal every twenty-four-hours identifying it as earth.

The British scientist, Sir W. Simens first discovered this mysterious energy signal in the 1800's. While touring the King's chamber Mr. Simens took a wine bottle and wrapped it with a wet newspaper turning it into an electric capacitor. The bottle gave off Sparks as it collected energy.

GOLD OF THE GODS

An old Alexandrine text makes particular mention of the weight of the Philosophers' stone-which it calls the Stone of Paradise. It states that: "When placed in the scales, the stone can outweigh its quantity of gold: but when it is transposed to dust, even a feather will tip the scales against it."

Everyone who has ever read about or seen pictures of the Great pyramids of Giza has asked himself or herself the question: 'Why were they built?' They were laid out according to the stars and aligned with true North with great accuracy. Once a person is aware that small viewing ports cut into the stone hundreds of feet long and aligned with certain star constellations and once a person knows that an inter-dimensional, super-conducting, white-powder capable of amplifying brain waves several billion times over was placed inside the pyramids then one has to come to the obvious conclusion that they are used as communication devices to teach mankind about the universe.

They were also built to warn mankind about cataclysmic events that could destroy all life on earth. What more important reason can there be for someone to build such monumental edifices than to warn future generations and make a phone call to the other side of the Galaxy? It is an inter-stellar communication device capable of communicating with thousands of intelligent races throughout the Galaxy. The communication was instantaneous and not dependent on the speed of light because the material in the Ark is inter-dimensional. What more important reason can there be than for an inter-stellar race to phone home? "ET phone home."

This raises the question: 'how many other worlds with intelligent races with pyramids are scattered throughout the Galaxy? The next question that comes to mind is: 'If were so smart then why aren't we building pyramids to communicate with other worlds? Why aren't we building monuments to warn future generations about impending cataclysmic events? The truth is we are not advanced enough to build a pyramid let alone figure out how it works and be able to predict when the next comet will hit. And we are definitely not organized enough to do anything about it. We haven't even reached the point where civilization was 12,500 years ago. Maybe in another fifty years or so if we don't blow ourselves up first in the meantime we might get it together enough to do something about our future.

The pyramids are just sitting there waiting for us to figure them out and take the next step, which is obviously to get in touch with our roots. Let us rise up from the tribal warfare over crude oil and get on with the business of becoming a member of the Galactic community, or as the American Indians call it: 'The Star Nation'.

CHAPTER EIGHTEEN

HOW TO LIVE 900 YEARS THE BLESSING WAY

In Old Testament times meals included a blessing over bread and wine (Gen. 14:18). Over the centuries the blessing assumed special importance, as can be seen in many sections of the Dead Sea Scrolls. Josephus writes about the Essenes: "They consider it a grave sin to eat or touch food before praying". In New Testament this prayer over food and wine (Grace) was converted into the symbolic ritual of eating and drinking the blood of a God. This ritual was "officially" introduced into Christian practice during the latter part of the fourth century.

Going back to pre-Egyptian times, the priest bread contained a large amount of the white-powder-of-gold. It was considered to be the food of the Gods therefore it was extremely important that you prayed over your food before eating it. Your thought waves would change the magnetic frequency of the ORME material thereby purifying it and making it more compatible for your body to digest. The prayer thereby making it more digestible changed the Messner frequency of the food itself. If you did not pray over your food it could make you sick. If harmful bacteria had come in contact with the ORME material or even the wheat in the bread when it was growing and you ate it without praying the harmful vibrations of the bacteria could be transmitted to the ORME atoms which in turn would make you sick. This is much like the water-molecule-memory-theory behind homeopathic medicine.

In homeopathic medicine the essence of a substance is diluted hundreds of times yet it still retains some of its properties. This is due to the ability of the water to retain past memory characteristics. Poisons diluted hundreds of times and deemed harmless by conventional science can still have an effect on the body. Combine a small amount of the white-powder-of-gold to direct the outcome or your medication and pray. Anything is possible.

LIFE EXTENSION

The rules of life extension are simple. Take care of yourself and don't subject your body to unnecessary risk. Eat things that are alive because they contain more super-oxidants, which mop up free radicals. Drink water that is charged with electricity. Rainwater is charged with high voltage electricity however you cannot drink it in most parts of the country due to air pollution. The upper atmosphere is charged with thousands of volts of electricity. When you take it into your body it increases the ionic flow of electricity throughout your body. Rainwater is a naturally charged subsistence. Anything you do to increase the natural flow of electricity will assist your cells during cell division. Living in Heaven or the upper atmosphere will increase this flow but at present we have no way to get there. See my book God Ships.

DNA AND CELL DIVISION

Try to imagine a flexible ladder one meter long composed of six billion steps. Each step is composed of two capitol Ts facing each other. The horizontal line of the T is made of 70% sugar and 30% phosphate. This vertical line is made of adenine or guanine, or thiamin and cytosine.

At the end of the T is a hydrogen atom. Hydrogen is composed of one electron and one proton. At the end of the phosphate rod the electron spins clockwise. At the end of the sugar rod the electron spins counterclockwise. The electrons of the hydrogen atoms at the end of the adenine and thiamin rod spin clockwise. The hydrogen at the end of the guanine and cytosine spin counterclockwise. This type of T is called a base or nucleotide.

Try to imagine twelve billion nucleotides facing each other in a strait line and then twisted a hundred thousand times to form a double helix known as deoxyribonucleic acid. This is DNA. Every one of your 6.3 trillion cells has one of these.

To create new cell the DNA must form an exact replica of itself. To accomplish this the DNA must unwind and split itself down the middle. The two must be exactly alike and in the same sequence. To accomplish this the cell must create another 12 billion nucleotides from food that is inside the cell. The presence of a few ORME atoms or manna allows the cell to transmute any elements that may be lacking. The absence of ORME material can cause genetic defects or even cancer.

Our body cells contain at least eight different kinds of virus. There are four RNA types and four DNA types. These viruses attract each other and pair up to form a bipole. To protect themselves against magnetic and electric forces they exist in a crystal form except when they are in active status they cover each other with protein. Even while they are in this status they act like crystals. Which are affected by sound waves.

When subjected to sound up to five megacycles they continually strain against each other, which produce's energy using the piezoelectric effect. When subjected to frequencies over five megacycles an inactive virus will disintegrate. Human DNA is tuned to frequencies between 375 and 385 megacycles (H.A.R.P. frequency) and also is under the influence of charged ions that travel through the nervous system. The action of the two forces causes the DNA to emit sound in the 1.9 to 2-megacycle range. The DNA uses this sound to detect what type of protein is missing in the cell. The sound maintains the virus activity in bipole form and also scans for protein.

The DNA and RNA virus protect themselves by covering themselves with phosphate and or sugar.

The flexible DNA ladder composed of six billion steps is twisted thousands of times so that for every turn of the coil there are only a few steps, which are called base pairs per turn. Every base pair has a different polarity, which was encoded by past generations.

Every time electricity flows through the nervous system, the DNA vibrates back and forth as much as one base pair. The more base pairs per turn the less the DNA has to move and the less energy it requires to reproduce itself. If there is more energy in the H bond atoms and many numbers of base pairs per turn then the DNA doesn't have to work as hard to reproduce itself. If you add energy in the form of electric frequency or charged ions flowing through the nervous system then the DNA will work flawlessly. This is what the manna does.

In the beginning the embryo cell in the mother's womb has 46 base pairs per turn. The H bonds are weak and the charged energy from the mother is strong stimulating the DNA to reproduce one time per second.

In the middle of the first month the number of base pairs has been reduced to 34 and the kinetic energy of the hydrogen bonds increases and the charged ions coming from the mother are unchanged. This reduces the rate of cellular reproduction to one a minute.

After the baby is born its cells have reproduced more than six billion times. At the age of two the baby only has 22 base pairs per turn and ten at the age of 35. By the time you reach age fifty the number of base pairs per turn of your DNA is only 10. After that the kinetic energy of the H-bond is quite weak and your cells don't reproduce as well. The faster you start loosing body cells the faster you age. Loss of body cells causes aging.

In order to slow the aging process you have to stretch the DNA to increase the number of base pairs per turn back to ten or more. There are many ways of doing this. One way is to go into space and live in the outer reaches of the ionosphere. Another way might be to wear magnets of the proper polarity on your hands and feet. Still another way might be to live in a mountain area with more ions flowing through your body.

I occasionally eat a teaspoon of the magnetic sand from off my beach. If nothing else it will set off the security alarms at the airport. The manna or ORME atoms can increase the ionic flow dramatically. It is up to you to choose your own way of winding up your own DNA. Pray over your food and choose wisely.

To live to five hundred and look good takes something with a little more kick; the gold of the Gods is what is needed. The correct procedure to take the stuff can be accessed by logging on to David Hudson's web page listed in Chapter one.

MORE EVIDENCE OF A PILL OF IMMORTALITY

The ancient Chinese Alchemist Wei Po-Yang wrote that the "Pill of Immortality" is made of Haun Tan "the complexion becomes rejuvenated, hoary hair regains its blackness, and new teeth grow where fallen ones used to be. If and old man, he will once more become youth; if an old woman, she will regain her maidenhood." Hindus worked with mercury burning it into the correct edible white powder by heating it to convert to gold.

LONGEVITY

Scientists have found that the longest living people on Earth who live to 120 and beyond live in high altitude regions that have a lot of minerals in the soil. They drink murky, glacier-silt, laden water filled with minerals and eat a restricted diet of yogurt and whole grains grown on their mineral-rich soil. It is the monatomic minerals in their food and water that contribute to their longevity. The mountainous terrain and high altitude guarantees that they exercise the large mussel groups in the legs and this in turn guaranteeing the heart will be strong. As long as they lead a strenuous life-style, eat well and take monatomic mineral anyone can live to 120 and possibly beyond.

Artephis, an alchemist of the twelfth century, claimed in his book: The Secret Book that: "...he had lived for the space of a thousand years, or thereabouts which has now passed over my head, since the time I was born to this day, though the alone goodness of God Almighty, by the use of this wonderful quintessence."

A French alchemist by the name of Fulcanelli tells in his book: Le mystere des Cathedrales how the secrets of alchemy were carved and concealed in the architecture of the great cathedrals. He is believed to have discovered the philosopher's stone by his study of the great cathedrals. The substance is an edible white powder or oil that imparts healing and spiritual benefit.

COLLOIDAL SILVER

I advocate the use of colloidal silver and other colloidal minerals. The Food and Drug Administration does not regulate colloidal silver because it was used prior to 1934. The medicinal use of silver goes back to the ancient Greeks who lined their cooking pots with it. The Pilgrims put a silver dollar in a gallon of milk to keep it from spoiling without refrigeration.

Mark Metcalf's excellent book can be ordered by calling toll free (888) 505-6005. I ordered his DC colloidal silver generator for $79.95. The book is free with an order. I figure I am at the age where it is time to start killing parasites. The older you get the more parasites build up in your body and the more your immune system is compromised. It finally gets to the point where vital organs like the liver and kidneys start to fail because they are overloaded with toxins.

Colloidal silver has been proven to kill over 650 different types of bacteria and some virus. There is no danger of heavy metal build up because the human body has the ability to process tiny atoms of colloidal silver.

The Environmental Protection Agency's Poison Control Center reports a No Toxicity listing for colloidal silver. In fact, it appears that harmlessness is one of the attributes of the colloidal physiology, regardless of content. "Many strains of pathogenic microbes-viruses, fungi, bacteria or any other single-celled pathogen resistant to other antibiotics are killed on contact by colloidal silver, and are unable to mutate. However, it does not harm tissue-cell enzymes and friendly bacteria."

Mark writes: "My next experiment was a little different. I cut flowers in the back yard and left one on a shelf without water for twenty-four hours. When I picked it up the next day it was completely limp. I made a fresh cut at the bottom of the stem and placed it in a glass of high concentrate colloidal silver. Each day it got better. On the third day, the stem had become firm again, as if it had just been cut. Not a single petal was lost. The second flower that was cut had been placed immediately in ordinary water. Many of its petals had already fallen. Through the flower was in water from the start, it was already dying.

It's as if there is something in silver that is tied to the very core of the life process itself. The noted bio-medical researcher from Syracuse University, and author of the Body Electric and Cross Currents, Dr. Robert O. Becker, MD, agrees. Writing about his experience with older patients, Dr. Becker wrote, "Silver did more than kill disease causing organisms. It promotes major growth of bone, and accelerated the healing of injured tissues by over 50%." He also discovered that silver: "profoundly stimulates healing in skin and other soft tissues in a way unlike any known natural process...." Silver is a conductor of electricity. Very small grains of silver speed up the healing process because they make it easier for the cells to transmit information.

Dr. Becker discovered that the silver was promoting a new kind of cell growth, which looked like the cells of children! "These cell grew fast," he wrote, "producing a diverse and surprising assortment of primitive cell forms able to multiply at a great rate, then differentiate into the specific sell of an organ or tissue that has been injured, even in patients over fifty years old."

The healing properties of silver are so all-encompassing that we see researchers expressing amazement time and time again. Alfred Searle, founder of the pharmaceutical conglomerate, wrote in 1919: "Applying colloidal silver to human subjects has been done in a large number of cases with astonishingly successful results. For internal administration, orally or by hypodermically it has the advantage of being rapidly fatal to parasites without toxic action on its host. It is quite stable."

To make a long story short colloidal silver solution can be applied directly on wounds that won't heal and dropped directly into the eye for an eye infection. The plaque on your teeth will disappear after drinking the solution for several weeks. NASA uses a silver water purification system for the space shuttle and so do the soviets. Japanese firms even remove cyanide and nitric oxide from the air with silver. Silver is the safest and most powerful medicine on earth. Silver kept the blue-blood families of Europe alive and healthy during the plagues of the middle ages. More information is available by calling the toll free number above.

Enclosed is a copy of a device, which imparts ions into water. It sells retail for $149.95. The address is:

www.hiddencures.com
E-Mail: Info@hiddencures.com

CHAPTER NINETEEN

CANCER CURES WHERE TO LOOK FOR THE GOLD OF THE GODS

I altered my thinking 180 degrees with regard to the anti-gravity, stealth atoms (ORME material). To my knowledge no one has ever looked for anti-gravity material on earth.

We know that ORME material exists and that half of the periodic table of elements can exist in the ORME state. We know that when in this state it has anti-gravity properties depending on temperature or static charge and we know that these atoms are created by temperatures over 5000 degrees. We also know that when volcanoes erupt, some of them spew out millions of tons of hot mud that comes from thousands of feet underground. Magmatic bubbles cook and rise up within the earth's crust much like a lava lamp only it happens over very long periods of time, millions, even billions of years. We know that when volcanoes do erupt they throw out light rocks like pumice and we know that certain volcanic mud is rich in the lighter minerals such as aluminum, selenium, magnesium, and lithium. Wouldn't it be logical to look for anti-gravity material in volcanic soil and volcanic mud? We know that in Tuxedni Bay the black iridium beach sands originate in this volcanic mud.

Over long periods of geologic time, where temperatures reach 1200 to 5000 degrees we can assume that stealth atoms are created. Billions of years later the lighter material would tend to float upwards. It seems logical to me that directly above this layer of magma would be a high concentration of anti-gravity material. We know that the lighter elements would float upwards as they do when different kinds of brachia, granite, and feldspar are created.

Could this theory account for vortexes on the surface of the earth and places of gravitational anomalies? The concept opens up a whole new area of exploration. If I can get a few people out in the back yard turning over rocks looking for anti-gravity material I will feel that I have done my job.

SCIENTIFIC AMERICAN

A December 8, 1989 article in Scientific American article documents the existence of microclusters. "Divide and subdivide a solid and the trace of its solidity fade away one by one, to be replaced by characteristics that are not those of liquids or gases. They belong, instead to a new phase of matter, the micro cluster. Micro clusters consists of tiny aggregates comprising from 2 to several hundred atoms." What we found out is that every element has a minimum size; it totally breaks up to on its own. Every element is different. Iridium has a nine-atom cluster platinum is a 5-atom cluster. Palladium has a 5 to 7 atom cluster and gold has a 2-atom cluster. Anything larger than that stays metallic and will aggregate and become more metallic. Anything less than that will literally break up on its own."

The actual process of breaking down rocks into its smallest component can be seen on any sand beach during a storm. The ORME atoms in a macro state when dried look like a white powder. If you look at them under a microscope they look like glass. When you heat the white powder to 1160 degrees in a vacuum it forms a glass like window glass. The atoms are too far apart to have any chemistry yet they are sitting at a distance and resonating in perfect unison. The atoms are rolling around and round forever and never run down because they are dipping into the zero point energy each time.

The last ten years of Bristol-Myers-Squib research shows that this material interacts with DNA, correcting the DNA. All the carcinogenic damage, all the radiation damage, all is corrected from the elements in the presence of the cell. They don't chemically react with it; they just correct the DNA curing aids and cancer on the atomic level. This is truly the light of life.

We know that plants absorb ORME atoms and we know it is needed for cell division. It seems logical that certain kinds of plants like the mustard, which is known to collect gold,

would collect more ORME material than other plants. Why not process the plants that contain the most monatomic elements and sell them as food supplements? To my knowledge no research has ever been done in these areas of food production. We have a food factory. Why not try concentrating mustard or spinach into a capsule and see the affect on the human body?

DAVID HUDSON

"...After I filed the 22 patents in 1988 on Orme Gold, Orme Platinum, Orme Palladium...(You could just as easily say 'ghost gold', 'ghost platinum', etc.... about 1990, my Uncle showed me a Time Life book on secrets of the alchemists. I put it off because I thought there was nothing in alchemy that could be of use to me. He kept after me and showed me a section where it talks about a white powder made from gold. The goal of the alchemist was to make a white powder of gold, which would serve as the container of the light of life. If you stand in it's presence, you don't age, if you partake of it, you live forever. So I begrudgingly read the book and have now read 500 to 600 books on alchemy and its history.

All of it goes back to a man the Hebrews called Enoch. The Egyptians called Thoth, and in Greece, they called him Hermes Trimesgritus, all the same man. It is claimed he ascended by partaking of the white drops, the man who never died, he ascended because he was perfect. The Bristol-Myers-Squib researchers have found that the precious elements inter-react with the cell by a vibrational frequency or by a light transfer to correct the DNA. It perfects the cells of our bodies. The element going into our body is not a metal, the element is an ELEMENT, and there is no heavy metal poisoning.

You can eat any amount of this white flower you want and it won't hurt you, it goes through your digestive system. In fact we took some brain tissue from a pig and a cow and we analyzed it, first we destroyed all the organic matter and did a metals analysis. Over 5% of the brain tissue by dry matter weight is RHODIUM and IRRIDIUM and no one knows it, because it can't be directly measured. The elements are flowing the light of life in your body; the elements are what the light is.... (Remember, this is ORME-RHODIUM and ORME IRIDIUM, detected by the 386-second spectrographic analysis).

I can show you four papers by the US Naval Research Facility where they have PROVEN that the cells communicate with each other by a process identical to superconductivity, but they can't figure out what is superconducting, It is these stealth atoms at work. The atoms are in your body, the atoms the flow the light of life and nobody knows they are there because they don't identify by (normal) instrumental analysis. The AURA is the MEISSNER FIELD from the superconductivity.

In our body we have what is called 'junk DNA'. There are over 30 aspects of DNA, which no one can figure out what they are there for. We only use 15% of our brain, what is the other 85% for? Did we evolve a brain we don't use? It's as if we existed at one time in a higher state of enlightenment and we have fallen to the state we exist in right now. There are still many mystery areas within living tissues.

There is some material from ancient Egypt that provides some insight into these white-powder; Budge calls it the Egyptian Book of the Dead and the Papyrus of ANI. This is the oldest Book of the Dead, found about 3500 BC in the tomb of Pepi the 2nd. It says: I am purified of all imperfections, what is it? I come before my father in heaven, what is it? I pass by the immortals without dying, what is it? The Hebrews worked in Egypt for many generations; they were the artisans and metallurgists. When they left Egypt, Bezalial, the goldsmith prepared the bread for the presence of God. Bezalial prepared this bread that the high priest took of, the millets. The word in Hebrew that literally means, "What is it?" It's Manna. The word Manna literally translates verbatim to a question: "What is it?" Look at the travels of Josephus and you will see this is so. The very same words were used in Old Kingdom Egypt in 3500 BC. Basically, these elements are naturally in your body, primarily rhodium and iridium, but gold (OR), means gold or the highest light.

The Bible says that Moses told the Hebrew people that they had not kept the Covenant, so the Manna was to be taken away from them, but will come back in the end times when we would be a nation of high priests, not an elect high priesthood. This is the food; this is the light you take into

your body. In fact if you ask a Rabbi if he ever heard of the white-powder-of-gold, he says: "yes, we know of the white- powder-of-gold, but to our knowledge no one knows how to make it since the destruction of the first Temple, the Temple of Solomon." This knowledge was not completely lost, the high priest who left the temple when it was destroyed went out into the desert and organized the community known as Qumrun, and they were the Essenes. In the copies of the Dead Sea Scrolls Uncovered, translated by Eiseman and Wise, this was known as GOLDEN TEAR from the EYE OF HORUS. It was the white powder of gold mixed in water. It was called that which issues from the father in Heaven.

You have to understand the symbolism, the understanding of; prepare yourself for the coming to the father in Heaven. To be inseminated by this father in Heaven in the bridal chamber, to totally be regenerated, to be purified, to be cleansed. Every cell in your body will be taken beck to the state it is supposed to be, when you were a teenager or a child. It perfects the DNA, and closes the light within the body until you literally reach a point where the light body exceeds the physical body. You literally light up a room when you walk in. The gifts that go with this are perfect telepathy, you can know good and evil when it is in the room with you, you can project your thoughts into someone else's mind, you can levitate, you can walk on water, because it is flowing so much light in you, you literally don't attract to gravity.

When you understand that your body can now exclude all magnetic and other fields, including gravity, you are no longer of this space-time, you become a fifth dimensional being. You literally can think where you'd like to be and you can go there and disappear from here. The ancient texts specify the appearance of other attributes such as healing with the laying on of hands, they claim you can resurrect the dead within two or three days after they die. Instead of using electrical paddles you embrace their people and bring energy and life back into them. Most groups don't receive this very well. It works.

Christ said to his disciples, "Don't touch me, I don't have on my earthly garments." They said, when will we see you again? Christ replied, "When you have prepared the proper food and when you have on your proper garments." What is the proper food? It is the food of the angels, the food of the gods, the Manna, the 'what is it'? The proper garment is your garment of glory, your rainbow garment, your Meissner field is what science calls it, and literally it's about 1000 times what you have now.

The amazing thing about superconductors they don't have to touch for current to flow. Electricity has to touch for current to flow but superconductors can sit at a distance from each other, and as long as they are resonant to each other they are AS ONE. They function as one. When you have your perfect super-conduction body you are no longer of this space-time. You are a light being, your mind is one with other people's minds. You literally know their thoughts and they know your thoughts. You're one mind and one heart and this is science.

The Bible says that the man who will plant the golden tree of life, which in Hebrew is the ORME tree, the name of my patent, and I had no idea of this when I filed my patent. My cousin joined the Mormon Church and they told her to do our genealogy. My thrice great grandmother is Hanna Disguise, the daughter of Christopher Disguise, brother of Claude Disguise, who if you read the book, Holy Blood Holy Grail, you know who the Disguise' are. Nostradamus worked with the Disguise family and Nostradamus prophesied that by 1999, the occult gold will be known by science. The old enemy of religion and philosophy is science. But in fact, science will serve up the confirmation and science will be the one to bring this to the world.

In Revelations it says: blessed shall be the man who will overcome, for he shall be given the hidden manna, the white stone of purest kind upon which will be written a new name, you will not be the same person. It's encoded in your DNA waiting to be activated.

Solar flares, which produce cosmic ray bursts, increase ionic flow through the body. These may actually be responsible for unlocking ancient symbols encoded in our DNA. Nature works in mysterious ways.

CHAPTER TWENTY

ETHICAL AND PHILOSOPHICAL CONSIDERATIONS

Enlightenment of the human species opens up a whole new chapter on morals and ethics. Is mankind ready to receive the responsibility of God-like powers? Some of the Gods of the past enjoyed inflicting pain and suffering, and they required animal and human sacrifices of their subjects. Other Gods were benevolent and kind, wanting only to assist mankind in its evolutionary quest for higher enlightenment.

We have read our history of the wrath of God in the form of bloody wars pitting one group against another as a kind of sport. The question is will we act the same way when we evolve to the god-like state? I can only say we have a long way to go but with a little higher enlightenment we might help make a relatively uneventful transition. Remember even the gods had problems.

INTERVIEW OF SOMEONE WHO INGESTED WHITE-POWDER-OF-GOLD.

The following transcript is from a conversation between Binga and one of the key players in the area of monatomic elements. The interview took place July 27, 1996. Prior to the recorded conversation, it was agreed that the interviewee was to remain as anonymous as possible therefore I cannot provide any specific information (e.g. name, contact information, position, location etc.)

For the record, I personally attest that this person is indeed both qualified and capable of speaking on the following subjects. This material is not copyrighted and may be reproduced and distributed freely, but only in its entirety.

"Strength in wisdom." <u>Binga@zz.com</u>
{I understand that you have ingested monatomics in the past, what product did you take?}
A combinations of Iridium and Rhodium.

{When?} About two years ago.
{How much?}I started with 250 mg a day and by the time I ended a 42 day fast was up to about 2500 mg. I probably took 2500 mg for a total of five or six days though, and I only did 250 mg three or four days. {Then went up to 100mg?} I'm impatient, I moved up very, very quickly. I started taking the material eight days into the fast. I did a lot of cleansing both before and during (slightly into the fast). During the fast, we monitored every single detail we could including: blood sugar, blood pressure, weight, temperature...things we could get a handle on.

{So you were fasting for seven days. On the eight day, you started taking 250 mg. Are you still taking it?} I still take material but I don't take it in that volume. Now you have to understand, these materials are present in carrots. They're present in all kinds of food.

{Yea, grape juice, bilberry, slippery elm bark...are you currently taking any isolated material?} Occasionally. But not at 2500 mg doses. {Why did you take it in the first place?} There were a lot of things we wanted to find out about and there was no other way to do it. I'll tell you right now: we didn't know if I would die or not.

{But you were willing.} I have no fear about anything like that. I really do not have any fear.

{What effects can you attribute to your ingestion of this material?} It wasn't very long after I started that the sound started occurring. The sound outside of sound That really is key to what can happen. Most people will hear that and think their ears are ringing. If you're careful, you'll realize that it's not in your ears. As you proceed with this, you'll realize that it actually moves outside your head and just above the crown of your head. It's more than a sound, it becomes an emanation...it really does.

And that's when you have something that you can work with. By far, the majority of people who've ever had to deal with this didn't have a clue. If they'd studied some Taoist alchemy...if they had studied anything and applied it to themselves rather than keeping everything outside themselves... they'd have known that the dialogue they build with that phenomena gives them the keys to the nest steps.

There really is a trade-off that starts to occur. That's what made my experience so utterly different from everyone else's.

{So you were prepared for it?}
No, but That a two day conversation in itself. If you want to get to the rest of your questions...it got to the point where I was standing in an electric flame. I could tell you a million things that occurred that were all way different from anything normal, but they occurred progressively as a result of me dealing with them.

{Where you psychic prior to ingesting it?}
Yea, probably. But what happened afterwards is truly unique. It changed everything in a way that hasn't made my life very nice. If you were to talk with some people, they would say it's made me impossible to be around...there's no way in the world to lie. It's nor fun. Most people are not going to be very alone. These material and now they've been dealt with historically have a rich, beautiful tradition built around them. Only now do I understand why.

{Any adverse effects?}
Well, my life is not Father Knows Best. {Are there any lasting manifestations?} Absolutely yes!

{Care to elaborate?}
Well. Psychic, physical, emotional! {Did you get more psychic?} Yes, oh God yes! Little things, all the time! It got to the point where you just didn't want to deal with it anymore. Like answering someone way before they ask you...it scares people. When this happens thirty times or more, they realize that it's not just a coincidence.

Do you think that your experience is typical of what others might experience?

No, I don't think my experience it typical of what anyone else experienced who did any meaningful amount of it. I'll tell you right now, I'm the only one who did what I did. All these sages that came out of the woodwork: after about 15 to 20 days, couldn't even make their fast. They started cutting corners and then said: "Well it didn't happen to me."

{Had you a previous kundalini experience?}
Yes, I had on several occasions in my life.

{Where you on the verge of having these experiences on your own volition?}
Well, its like this: I knew what to expect. When it started to occur, it didn't frighten me. It's like waking up in an Astral dream, the first few times you frighten yourself into waking. After that you finally say 'Wait a minute ' don't do a thing I think we have finally hit the trigger, and then you float out. That's something I had worked at for years, and years, and years. I got so far into it, I finally quit doing it. It got to where I could do it any time I wanted to within five minutes or so.

{Out of body experiences?}
Yea, I could put myself right there.

{Well you've been on a spiritual path since what" Age eight?} My whole body has been tied to it. Being down-to-earth, incredibly well-grounded, reasonable, logical and not giving up common sense in very important in this work. We're going to have to call it a prerequisite.

{What do you think the mechanism is at play here? Is this a chemical reaction?}
Well, everything in your body, we could sooner of later get around to saying, is chemical but the junctions in your nervous system are all made up of these materials. You've probably seen the paper—Superconductive Tunneling and Biological systems. When you start overloading everything with this material, your real potential begins to be accessed. Now that's one thing hat happens. The real trick to it deal with the monatomic gold in a perfectly charged way. We could go on forever on this...

{We have a lot of ground to cover too. Has anyone taken monatomic gold?}

Yes. {Who? You?} Yes. {How long did you take it?}

Very little. Not for very long.

{Why did you stop?} There's very little of it to go around.

{You would have taken more?}
Of course I would have. I'll tell you right now. Monatomic gold can only be typified as the precursor.

{So you've taken as much as you could get?}
It's not a matter of how much; it's a matter of having it prepared properly. It takes very little when it's prepared properly.

{What do you think a full course of the gold for humans is?}

A full course would be one application.
{Like 500 mg?}

No, I doubt it's anywhere close to that. 500 mg, when you throw it our on a table, isn't very much. But, we're not talking volume here. Volume has nothing to do with it. It's also what you bring to the table that will have a great deal to do with the banquet you enjoy.

{What role do you play in all of this?}

Well, my roll is my own. It's not really connected to anything. But, it is important to allow people a little easier access to this body of knowledge that they have had historically with alchemical information. I used to be so angry about how the alchemical texts were written because so much of it is misleading. Some of it is bad. Some of it's misinformation.

{We all are aware of the philosophical claims made for these materials: do you think the claims can be borne out?} Absolutely!

{Will humans truly affect bio-location?}

Absolutely!

{Any you just take along your physical body?}

Absolutely! Absolutely!

{It's not a purified physical body? Sort of a less dense version of what we have?}

Listen it is physical. It is so physical you can eat, have sex, bleed.

{Is this the voice of experience?}

No, But I'm telling you it's absolutely physical. It's not allegorical. It's not symbolic. Those are nothing more than Tinkerbell and Disney.{Could we beam ourselves to an inhospitable? environment and die?} No you can't. {It's physical?} It's physical here where it's supposed to be this way. {So you become whatever is appropriate for the place you beam to?} Yea, of course. You'd hate to show up at them all in you painting cloths. But from the point of logic you are using right now, you can only apply that logic to show that might be. When it occurs, You've already moved yourself to that place. The change takes place at the same time.

{But you still have memories of this third dimensional reality and your previous life. You are still you?}

It's not just a memory. Which do you think would be the most amazing miracle: it I brought someone back to life, or changed water to wine? The truth is: they're both the same. If you can do one the other is just as possible.

{It's just moving atoms around?}
It's even simpler than that. You're turning, literally turning, ninety degrees. There's another reality right around the corner. And when I say right, I mean ninety degrees. It's physically right around the corner.

{Isn't it just a change of prospective or perception?}
No, there's a whole new deal there that is beautiful and so totally 'other' that it's nearly impossible to imagine from here. But you can go both ways. Once you are capable of one, you can do the other. Like I said, water into wine is no different than turning lead into gold.

{Or turning the sky green?}
That's right, once you learn to cheat this way you can cheat any way you want.

{How about the technology claims? Like the fuel cell application?}
Absolutely! All the technological and biological possibilities are incredibly real possibilities.

{Do you see any other uses that haven't been mentioned?}
Thousands... Is there anything that concerns itself with cataleptics?

{Are we finally going to get off the grid? Will we be free?}
Well we may be, and there will be many, many, many hundreds and thousands that will. But if you're talking about the five billion or six billion: most of those are going to perish. But that's OK, it took this many to get us where we are now.

{So we're taking 144,000?}
Well, don't hold me to a figure like that. There's just a logistical problem that goes along with this. You can't just hand it out in a crowd, there's a lot more that goes into this. But let me tell you the capabilities of discernment that go along with what I'm talking about are great.

PART 2 SUBJECT FOR THE RECORD - INTERVIEW 2/2 FROM *BINGA*

{So when you're talking about 'this stuff' do you mean Rhodium, Iridium, Gold? Do you mean all of it?}

No.

{You only mean gold?}
Yea, that's what I'm talking about.

{So what about the rest of the stuff? It is only gold that will do the philosophical?}

No, the other will do a lot of things but I'm talking about something beyond most people's ability to imagine.

{You said before that you thought many people might die from the ingestion of these materials.}

The emotional shock of it would be enough to kill most people. Look, even if you took someone three-quarter, brain-dead and started pumping it down them, in a certain amount of time you would probably start to affect them on an emotional level.

{Do they need to believe it's going to work?}

No, this doesn't require any belief. These are physical materials, they're not allegorical.

{What can you do to prepare yourself in order not to die or suffer any adverse effects}?

You cannot sit in a cloistered room and prepare yourself for any type of disturbance. Your ability to deal with being disturbed, having your boat rocked, putting up with the storm, being tough, being gentle, being loved, being harsh, being all of those things...

{Being real?}

Yes, these are the things that prepare you for this.

Thinking everything is love, beautiful and New Age horse sh, isn't going to do it. You're going to get slam-dunked.**

{What types of adverse effects do you anticipate?}

It could be as simple as someone with cancer, who doesn't believe that something so weird could possibly have an effect on them, taking it at the request of a loved on and finding themselves miraculously cured. Now that could be a huge emotional shock to someone. If you really think about it for a minute, people beliefs systems act as the best prisons in the world.

{I was speaking more in terms of physiology.}
No, no I don't think so. Your whole body depends on this just to be able to act like you're alive.

{Do you think everyone should ingest monatomics? Who should or should not and why?}

Everyone does ingest monatomics. They're present in common everyday foods. I don't think life would go on at all without them.

{In concentrated form?}

In concentrated forms, I don't think it is for everyone. I don't think many people, in terms of percentage of the population are going to be compelled to deal with this in the way someone who has spent years working on a spiritual path would be.

{So, only people who are working on a path ...}
Well, People have been drawn to it from some very deep part of their soul.

{With the many other materials purported to be monatomic on the market today and the prospect of Even more in the future, how will we be able to tell what is real and what is not?}

99.9% of all materials out there are total bullsh**. Most People are being sold a bunch of hype. There are some materials out there with monatomics in them, but generally you can get just as much from a couple of carrots. You know it makes me feel very bad that this is being capitalized on but such unscrupulous people.

{So how can we tell what is real?}
There are people selling materials that have mercury and lead in them. They know it, and they're not including it in the analysis they are putting forth. Through the analyses we've done, I can assure you that any 'white gold' products (and there are many out there), contain no monatomic gold. Some ar even contaminated with lead, a fact they choose not to include in their public analysis. Now, isn't that spiritual?

{Yeah, this brings out the best and the worst in people. So there's no way for people to tell if the material is safe, or real, or what?}

The only way to tell if a material is monatomic is through testing. If they are willing to pay for it. It can be done, but it's going to cost $3,500 to

$4,000 to do a certified analysis. There's only one lab, that I know of that can do it.

{Can't you put it aqua regia and if it doesn't dissolve....?}

Yes, that would be a quick and dirty test. But all you would know is that it's monatomic. You wouldn't know what it was: osmium, ruthenium... etc. But you could also have something left there that's not monatomic, especially if the source wasn't from a pure metal standard.

{For what purpose, in your opinion should the monatomic **gold be used?**}
Most people who are going to take it on purpose are doing so for the sake of expanded consciousness. The gold has unique physical properties. There are technological applications but those have never been applied. This is the edge of the razor here, can you feel it?

{What would you do with a pile of monatomic Ruthenium?}
Well, quite often those elements are very desirable in electronics.

{And Palladium?}
Well, those could be applied to a lot of catalytic applications.

{Osmium? Electronics! Iridium?}
M-state can be used in any type of surface-wear applications, plating, Where very durable corrosive resistance necessities are present and any catalytic application. The same is true for Rhodium. Of course the biological applications for those two are the best known. {You said previously that you don't have to be a chemist to make this stuff, what did you mean by that?} Any housewife in Dallas could make this material right now in her own kitchen, if she did her homework.

{But is still involves chemicals...correct?}
Yes, it does. Lots of study, lots of time, lots of diligent observation.

{Lots of holding your mouth right, saying the right things over it, putting your hands on it?}

No, you could do it in a few years probably. But it's one of those things that if you screw-up part of the way through it, you have to go back and start over. You can't imagine how disheartening that is. It really is an exercise in patience, perseverance and observation. My position on this is if someone id jazzed about it, they should bring it into their life. What they bring is a rich tradition that is full of guidance that will come to them when their heart is in the right place. {Some people have children, bills to pay, and Oprah to watch. They don't have the time to do it.} And that's a choice for them. This isn't something on the level where people just casually say they have an extra $100,000 or $500,000....there's no amount of money that can be put on what we're talking about.

{There are a lot of people who think you could.}
Those are the same people who should study the Mayan cargo system which still exists today. That is: you don't send money to put on the parade when it's your turn to carry the flag. You carry the flag because that's the only thing that counts for carrying the flag. When people think for a second that they can spend $500 and get a seat net to the Buddha they're wrong.

{What do you think the ramifications of the distribution, and subsequent ingestion, of monatomic will be on: human evolution, society, technology, finance, consciousness, the planet, the known/unknown universe?}

Everything we know will be changed...on all levels.

{Could you be more explicit? What do you think will happen to society?}

When people start taking it and the material start being applied, not just in terms of oral ingestion, the world will never be the same again.

{What is your vision of that?}
Everything from biology to power production, political structures (decentralization of government), everything has the potential to be radically changed. So many of the things I just mentioned are based on structures and paradigms that are centered around the control of resources....

{And perceptions?}
And perceptions and all of those can very easily be displayed by the applications of these materials.

{What do you think will be the effects on the known and unknown universe?}

They're going to trade places. The expansion will be so great that expansion is what it is.

{The process is what it's all about.}
Yea, it's not a matter of getting to a point where we've made this achievement... we've become

We have become aware because of the speed of this achievement that it's an expansive never-ending process.

{Are some of us are aware of this?}
Some are. It's balance through movement. Like movement you have in the European Tarot deck's two odd discs. It's like the movement of a bicycle, the movement forward is what allows for the balance. The stability is stability with no movement.

{Many people I come in contact with, think of it as a destination.}
There is a lot of belief that this is 'the end'.

{Do you see only a few people taking it at first?}
Only a few.

{Do you foresee a time when it will be available in vending machines?}
No, I don't. If you study how our planet was populated, creating the beings we are now, you'll find that political structures have always made this an elitist phenomenon.

What kind of time frame are we looking at? This is total conjecture on my part, but...a lot of this will remain in research for the next five years or so. After that, there may be efforts for it to trickle out in some way. There are other people, I personally know, who are so far down the road in applications and technologies (not necessarily in terms of production capabilities)...but you throw money at that one end and the problem goes away very quickly. There are many people working in this same project, as there have been in the past. There are several way of making this material.

You've said that there are different forms of monatomic Gold? So you mean different spin states?

It's not a matter of a higher spin state. It's a matter of once it is in the high-spin state, it being properly charged. That involves very subtle operations. Very subtle. We're talking beauty again, that's how subtle.

(*Author's note: This is why Christians pray over their food to bless it so it will nourish their bodies. The process charges the monatomic orbitally rearranged atoms.)

There 'is' a difference in their materials. They are all individual, unique material. They have their own qualities. That's why they have different names. To lump them all together under on heading as monatomic gold is just not accurate. There are different materials.

You are making a distinction between isolated monatomics and more highly charged monatomics?}

Yes, Yes I am. This aspect has not been addressed as far as I know.

{So what is the correct term to use in referring to the higher-charged materials?}
I call it the Philosopher's Stone. You could call it the Manna. There's a lot of things you could call it.

{But there's as much difference as night and day...ingesting this over the other?}
That's an understatement.

{Are you aware of other materials which carry the same attributes and implications?}
No, this is it...the real thing...the only thing.

{This meaning higher charged monatomics?}
You have to understand: the isolated materials are The Precursors. They were very difficult precursors to ever came up with. Very difficult. But it's the battery that you have to charge. It has to be charged to be able to

turn over the governors that are present in our genetic code. Those were programmed in.

{What practical advice do you have for those of us who know about these materials, and the potential they hold, but are living outside of ground-zero?}

Do your homework. Don't let anyone impose 'their' truth on you. Let them present the various facts and make your own decision. Keep searching for answers...you own answers. That seems a little patronizing in general, but realize that it takes work...physical work. this is not something you read into existence. You read enough to know that you have to do something physical. If all you do it read the book In pursuit of Gold-there's enough information there (if you apply yourself) to go very, very far. Foe someone who's doing nothing, that is a huge step forward.

{Do you have any specific advice for members of the *SOSF?}*
This material is real, but if someone expects to eat it and suddenly fly through the chimney: they're in for a real surprise. But the material, in terms of medical performance, is real. It will perform as expected and then some.

{And the technological part? Some people are expecting to make some money on their investment.} They very well may. But there are some many other uses for this material...

{If you had it all to do over again, what would you do differently?}
Absolutely nothing.

{No regrets on any of it?}
None at all.

ADDENDUM

Subject: pursuit of Gold Reference from: binga I had several requests for information about the book suggested in the 'For the Record' Interview.

Author: Lapidus Title: In Pursuit of Gold Alchemy in Theory & Practice Publisher: Woodstock

Beekman Publishers, Inc.

Strength in Wisdom, Binga.

CANCER CURES

I have heard of people curing cancer by drinking massive quantities of carrot juice. Carrot juice is high in monatomic rhodium. It wasn't the carrots that cured the cancer; it was the monatomic rhodium helping the body to heal itself by super-conducting electricity during cell division. Complete cell division cures defects in the DNA thus producing healthy cells which help a person get well.

Linus Pauling, two-time Nobel Prize winner and science leader, stated: "You can trace every sickness, every disease and every ailment to a mineral deficiency."

U.S. Senate Document 264: 74th Congress, 2nd Session, 1936 reported that depleted soils in America were raising crops enormously deficient of the mineral needed to maintain good health in the American people. Without minerals, vitamins are basically useless for health. Keep in mind this was 1936 when there were still a few minerals in the soil. "The Earth Summit Report" 1992 issue reported that the level of soil base minerals in North America was only 15% of what was found 100 years ago! And you wonder why cancer is growing by leaps and bounds in America for the last 100 years! Other degenerative diseases like heart attacks didn't exist in America until about 110 years ago. But don't worry! The FDA apparently wants to remove vitamins and mineral pills from the American diets!

Medical research in Sweden indicates that a trace of silver in your diet is very important for health. Apparently only colloidal silver can restore the silver that used to be in the normal diets before cancer got established in America.

Tom McCrorey, who had been told he was going to die of incurable, cancer in less than one month invented Black Salve, which is also called Balm of Gilead, back in 1890. He went on to live another 70 years as a very healthy, colorful man in life. Even back then the medical field had very little interest in a Black Salve that cured cancer. It threatened medical profits and so was ignored! An underground culture grew up where livestock infected with cancer was treated with this black salve and it apparently worked like a charm in curing livestock. In time humans started using it.

Rene Caisse, a nurse living in Canada discovered a cancer cure through a patient in the hospital she was working in. The patient had been cured of cancer by an herbal remedy given her by an Ojibway herbalist. Rene left the hospital in 1922 at age 33 and went to Bracebridge, Ontario, Canada where she began administering Essiac to all who cam to her. The majority of those she treated had been came on referral with letters from their physicians certifying they had incurable cancer and they had been given up by the medical profession as untreatable building lent to her by her patients. For a period of 60 years she administered Essiac both orally and by injection. In cases where there was severe damage to life-support, organs the patient died however they live much longer and pain free. Others listed as hopeless and terminal but without severe damage to life support organs were cured and lived another 35-45 years and many are still living.

"Unless we put medical freedom into the Constitution, the time will come when medicine will organize into an undercover dictatorship, to restrict the art of healing to one class of men, and deny equal privileges to others...all such laws are despotic and un-American, and have no place in a republic."

Dr. Bengamin Rush, Surgeon General of George Washington's armies, and signer of the Declaration of Independence (1787).

No law or pretended law of Congress and federal agencies is higher in authority than the natural law right of the American people to survive according to Thomas Jefferson, author of the Declaration of Independence which founded America as a nation in 1776. This natural law right to survive comes directly from God and no government on earth has authority to modify or cancel this natural law. John Locke and others who taught the basis of natural law clearly stated: "The highest legal right of mankind and higher in authority than any government on earth is the right of survival."

For the F.D.A or other federal agencies to dare claim they have the legal right to decree the mass murder of millions of Americans by denying them knowledge of cancer cures is wrong. Thomas Jefferson and other founding fathers of America would rage and denounce Washington, D.C. as a total barbarian government and outrageous tyranny for daring to try and mass murder millions of Americans by police state decrees of intended mass murder and pretended law (tyranny).

Tyranny is defined as a hostile to the God-given rights of mankind. The first two paragraphs of the Declaration of Independence of July 4, 1776 legally asserts that all authority for government comes from God who does not authorize any government on earth to become a tyranny against the people. If once a tyranny, it is no longer a valid or legal government in power.

I have no doubt that the human body can be transformed over time into a light being by the ingestion of the inter-dimensional material for a period of decades. Eventually you become what you eat. Light beings have the power to trans-locate existing at any place and in any form at will. Other powers include manifesting thoughts into third-dimensional reality. Entrepreneurs do this all the time.

SODIUM BURN

An electric ark will work to create the monatomic atoms from iridium sand however there is another way. It requires mixing an equal amount of sand with metallic sodium and allowing it to burn in a chamber filled with inert gas.

Afterwards the monatomic are precipitated out using alkaline. This is a rather expensive alchemy however it works. I have a lot of the black iridium sand. I can send you ten pounds of the stuff if you send me a check for $25 to my address.

I am finding quite a bit of ORME material in the Cook Inlet mud. Enclosed is a copy of a test of a sample of the black, beach-sand done by Bremner Technologies of Las Vegas, Nevada. They melt the material with huge electron beams much like in your TV set only much larger. It is done inside an argon-gas-filled, chamber to keep the ore from oxidizing. After it cools the resultant lump of glass is crushed and the metals are leached into solution. The elements are precipitated out of solution by changing the pH. When acid is added to change the pH it becomes visible in solution. It is filtered out and dried. The dry precipitant is melted with a flux to get some of the metal. It should be obvious by now that this process will not show the monatomics.

After that experiment I would like to try the non-magnetics that are left over after separation with a magnet to see if there is any difference. I need to do more testing and I could use grant or seed money to help me pay for some of the equipment. I want to separate the magnetic and have them tested along with the non-magnetics to determine where the values are located. I have the addresses of several good laboratories in New Mexico and Arizona that can detect the platinum group.

I know people look at me funny when I tell them that I am looking for anti-gravity material. I don't care because I know don't have all the information. Only the readers of this book will be able to understand that much of the ORME material is mixed up in water solutions and with other materials in the soil so it won't float away as suggested by some.

The whole idea of looking for anti-gravity material on earth still kind of warps my mind but the more I think about it the more logical it seems. The Gold of the Gods has given me a new direction in life and a goal unlike any other, a goal that could raise mankind to the next evolutionary level.

If only our leaders were just a little bit smarter we wouldn't be two hundred years behind in our intended evolutionary development. Too many of us think that mankind is far superior to all other life forms and he is the most intelligent creature on the planet. If mankind annihilates himself with disease, or poisons the environment, or bombs himself into oblivion he will be the dumbest creature on earth.

A TECHNOLOGY SO ADVANCED IT SEEMS LIKE MOVIE MAGIC

When describing magic is helpful to use examples seen in movies.

"The force" in Star Wars which the Jedi (Luke, Obi-wan, Darth Vader, etc.) can be utilized. Conformably, the name "Skywalker" (last names of Luke, is derived from 'loki' who is the Norse god of fire (which correlates with

Prometheus). References to psi abilities (telepathy, telekinesis, etc.) are also subtly in the movie 'Contact', one of which is in the scene where Jodi Foster (or the character she is playing) first hears the signal from outer space through her headphones which are connected to the huge parabolic antennas. In the scene, Jodi is quietly sitting with her eyes closed like Buddha in meditation, then the camera closes in on her left eye until the eye fills the whole screen, at which point her eye opens as she finally hers the signal from 'out there'. It is comparable to Buddha achieving enlightenment. It is a metaphorical depiction of opening up the third-eye-'psychic'/telepathic sense, we could say it shows activation of the telepathic 'antenna' in parables.... So the big "'parabolic' antennas" in the scene are really a pun!?

Anyway, the focus on the left eye relates to Horus' left eye which was lost in the battle with Seth; so Jodi opening the left eye symbolizes regaining the lost 'eyesight' -psi/psychic sense which is right-brain related (left eye is connected to the right brain, and the right eye is connected to the left brain). The metaphor is that when the psi sense is opened, it receives information just as the huge antennas hear radio signals. Eventually, after the scene in the movie, Jodie goes through an inter-dimensional travel to the Vega star system, which can also be taken to refer to the nature of psi - which is basically an inter-dimensional perception. It also recommend watching the Sci-Fi TV show, Babylon 5' for a great depiction of the nature of the 'telepaths", a relatively literal depiction of the 'big picture'.

The perpetuation of the 'psi' blood line (which the Catholic Church made every attempt to stamp out during the Middle Ages with the Inquisition) is alive and well today. Likely there are 'keepers' of the "Isis DNA", of that can be called that, and this is probably connected to the mystical secret societies that existed throuhout history who seem to have influence the course of history greatly from behind the scenes, probably both positively and negatively.

Through it is likely that the 'psi DNA' is not "active" in most people for a long time (since the flood?) as may be suggested by inactive state of Osiris, Saturn, kind Arthur/Merlin, and even Jesus, etc., and their destined 'return' in mythology. The resurfacing memory of the ancient past, the supposed underground civilization, and the supposed 'hibernating' beings in Mars, both again supposedly, beginning to come back our may all correspond to this also.

So, it could roughly be inferred that the 'Jupiter' force, the "Dark Father', the suppresser of the 'gift of fire' is to come back in 2008? Has our technology and awareness evolved to the point where we can begin to access the forbidden "knowledge' - the lost past, Atlantis, Mars, "magic", etc.? Use of m-state may bring it back.

SPIRITUAL PURIFICATION RITUALS

These rituals have nothing to do with religion. Religion is about preparing oneself to die. This book is not about dying. This book is about living on earth for a very, very long time and doing it the right way without sin. Spiritual purification is completely ignored by western civilization-possibly because it does not serve religion.

The decision to take m-state should not be taken lightly. It is important the information in this book be used only for good. If this book were to get into the wrong hands it could mean the end of civilization as we know it. I highly recommend that a person perform these spiritual purification rituals two or more times per day to cleanse one's soul before and after ingesting m-state.

What I mean by spiritual purification is not bowing to Mecca or saying Hail Mary's. Religions do not purify one's soul. Religions only promise a quick and easy shortcut into heaven through the forgiveness of sin. This usually requires that you chant a few prayers and give them a large sum of money. The goal of most religions is world domination through either conversion or war. When governments get together with religion the result is war and much bloodshed. This sort of behavior does not purify one's soul.

SPIRITUAL PURIFICATION RITUALS

1. Chant the name of your God day and night. It can be Jesus or any God, which does not violate your present religious conviction. OM NAMAHA SHIVAIYA is the traditional chant. Chanting out loud at first is necessary to form the habit. Eventually your mind

will automatically chant in silent mode. This ritual purifies your thoughts of evil because you eventually become one with God aware of his presence at all times.

2. The fire purification ritual requires that you stand in front of a bonfire twenty or more minutes a day. Several lit candles can be used instead of a bonfire or any large open flame of natural source will work. I wouldn't recommend gas or oil flames because of the carbon monoxide-polluted air they give off. The idea it to let you human aura circulate through the open flame which purifies and revitalizes it.

3. The water purification ritual requires that you immerse your entire body in clean rainwater or water from a flowing stream several times a day. Do not use tap water, treated swimming pool water or water from stagnant ponds, as the molecular vibration of these waters is detrimental to your aura. Swimming in the clean ocean is good as long as your beach it is not near industrial pollution. The water must have the correct electrical vibration to purify your body and spirit.

4. Last, but not least is the BREATH OF LIFE. Breathing correctly is absolutely essential to a healthy, long life. I recommend getting a Yogi instructor to show you how to do this correctly but for those of you who do not have access to this knowledge I will try to describe how it is done. Basically it is accomplished by breathing through the nose so that the organs and nerves in the sinus can absorb m-state which Eastern civilizations call Paranna energy. It has been proven that athletes can lift more and run faster after taking a few of these breaths. The breathing is done consciously twenty breaths where one merges the inhale with the exhale at the bottom of the breath. The breathing is done naturally and never forced. Experienced yogis can breathe through one nostril twenty breaths and then do twenty through the other nostril. This clears out both halves of the brain. For thousands of years the purification rites of Yogi and mantras, have been used to achieve

spiritual purification and life spans measured in hundreds and even thousands of years. Leonard Orr a western practitioner of yogi and a leader of the New Age Movement decided to find out for himself if there really were people living in the same body three hundred years and more.

Leonard Orr has published more than twenty books in a dozen languages including Rebirthing in the new Age; breath Awareness; Babajii, the Angel of the Lord; and Government Without Taxes. His international Rebirthing movement has served over ten million people on six continents.

On Leonard Orr's first trip to India in 1977 he didn't find any immortals but a female traveling companion did however. Babajii who is mentioned in the Bible as the Angel of the Lord is thousands of years old. Since then they met more than eight individuals who have met his minimum immortal criteria of 300 years in the same body.

Mr. Orr's book: Breaking The Death Habit, The Science of everlasting Life is a practical discussion of the basic ideas and practices for personal mastery and health. The book presents the minimum foundation for personal immortality, which Mr. Orr learned, from true immortal yogis. "God has given us the practical tools to be immortal. The bottleneck is dead orthodoxy's, philosophies, and beliefs, not only in churches, but in education, science, business, and government."

"Before I met Babajii in the flesh, I had figured out that breathing in cooperation with the mind was the key to the health of the body and mind. I had concluded that the breath of Life could be the "Fountain of Youth" and, therefore, the key to the eternal life of the body as well as the mind. Since Babajii has mastered the eternal life of the spirit, mind and body, his body, though thousands of years old, has the appearance and integrity of a young man." Those who seek him out at his ashram North of Deli, India will be rewarded.

The recitation of mantras is of extreme importance along with the other purification rituals of fire and water. " Om" or Aum" has been changed to "Omen" in Judaism and to "Amen" in Christianity. Some Moslems use this name also: "Om Allahhoya Om." The name reveals that all these major religions have their roots in the same God. If you study Exodus, Chapter 3 through 6, you will discover some fascinating details about the Eternal Name of God. Babajii appeared to Moses in the burning bush and gave him the eternal Name. This name is OM NAMAHA SHIVAIYA. You can still find it in Hebrew as Ya Vah or Ya Weh. Hebrew is spelled in reverse order from English and its written form has no vowels; thus "Ya Vah" is the last two syllables of Shivaiya in reverse order. The King James Version of the Bible translates it as Jehovah. In Hebrew today it is relatively common to say, "Ya Vah Shim Omen."

The religion of the Bible is the same as the religion of Shiva, the Sanskrit word for God. Sanskrit is the spiritual language of humanity. It Precedes the Tower of Babble. It is the language of immortal yogis."

When you study it you will find that all religions actually worship the same God. What is so difficult for people to understand that God can manifest himself or herself in physical form in hundreds or even thousands of human bodies at the same time? Such a God exists simultaneously, as a multidimensional spirit capable of materializing into the third dimension on any planes at will. If you seek him out his spirit will likewise exist in you.

CHAPTER TWENTY-ONE

HOW TO MAKE THE PHILOSOPHER'S STONE

In the late 1970s an Arizona farmer named David Hudson noticed some very strange material as he was doing some gold mining on his land. Hudson spent several million dollars of the following decade figuring out how to obtain and work with these strange materials.

In 1989 David Hudson was granted patents on these materials and methods for obtaining them. Two of these patents can be found on the Web.

http://monatomic.earth.com/david-hudson/patent-us1.html
is Hudson's British Patent.

http://montaomic.earth.com/david-hudson/patent-oz.html
is Hudson's Australian Patent.

During the early 1990s Hudson toured the United States giving lectures and workshops about what he had found. Transcripts of portions of three of David Hudson's lectures are available on the Web .The most complete of these transcripts is the transcript of his Dallas lecture and workshop. You can find this transcript on the Web at:

http://monatomic.earth.com/david-hudson/1995-02-dallas-toc.html

Two other Hudson transcripts can be found at:
http://www.cris.com/~Notnorml/alchemy1.html

Hudson's Portland lecture.

http://www.eskimo.com/~billb/freenrg/hudson.txt
http://elaine.telport.com/!boydroid/gold.html is

Hudson's Virginia Beach Lecture.

DID HUDSON FIND THE PHILOSOPHER'S STONE?

Since ancient Egyptian times, alchemists have worked in secret to produce something called the Philosopher's Stone, or the Elixir of Life. The material has been called ORMEs, monatomic gold white gold, white-powder-of-gold, ORMUS, m-state, AuM, microclusters, and manna.

David Hudson calls the material he found Orbitally Rearranged Monatomic Elements or ORMEs. He refers to them as monatomic elements in a high-spin state.

Since Hudson has patented his process for obtaining and identifying these elements, we would like to suggest that the terms ORMUS and M-state be used when referring to this state of matter.

The ORMUS or M-state material is thought to be the precious metal elements in a different atomic state. The following elements have been identified in this different state of matter (there elements, with the exception of mercury, are listed in Hudson's patents): Cobalt 27, Nickel 28, Copper 29, Ruthenium 44, Rhodium 45, Palladium 46, Silver 47, Osmium 76, Iridium 77, Platinum 78, Gold 79, Mercury 80.

All of these elements are abundant in sea water.

According to David Hudson's discoveries, these elements in the m-state may be as much as 10,000 times more abundant than their metallic counterparts. There also may be other elements which occur naturally in the m-state. Various researchers, working independently, have identified there materials in this different state of matter. They have arrived at many of the same observations.

These m-state elements have been observed to exhibit superconductivity, superfluidity, Josephson tunneling and magnetic levitation. It looks like there are an entirely new class of materials. There m-state elements are also present in many biological systems, They may enhance energy flow in the microtubules in side every living cell.

It appears that this state of certain of these elements have been known throughout history. Several of the procedures for extraction or making ORMUS have been adapted from ancient alchemical texts. We believe that the Philosopher's Stone and the Biblical manna are both variations of this state of matter.

Some recommended alchemical texts related to the Philosopher's Stone are "Sacred Science" by R.A. Schwaller De Lubicz and "Le Mystere des Cathedrals" by Fulcanelli, are available from Amazon.com. Another source is "occult Chemistry" by Leadbeater and Besant. The premier treatise on the subject may be "The Secret Book" by Artephius which is available at this Web site:
http://www.levity.co/alchemy/artephiu.html

There may be several paths to the philosopher's Stone. There may even be several different Stones. More research on the nature of M-state is needed. Since the ORMUS materials are much more prevalent in nature than their metallic counterparts, they can be extracted with some time, effort, and understanding. We invite others to join in the quest for knowledge of these materials.

THEORIES FROM PHYSICS AND CHEMISTRY
ABOUT THE *M-STATE ELEMENTS*

The following information is presented to promote scientific research into the nature of these materials. Although there theories are based on our best knowledge at this time, further scientific research may prove some of these theories to be inaccurate. Remember that the following are just theories.

* Anomalous responses to gravity
* Superfluidity
* "Tunneling" through solid objects
* Deformed nuclei in a high-spin state

I can see where this material repels gravity and jumps away from a magnet. The physical shape of the particles is particle is nucleus with a ring of electrons. Whenever you have a ring of electrons going around at the speed of light or greater and it encounters a magnetic field the particle will produce a magnetic and gravity field of the opposite polarity.

"Tunneling" through solid objects can be explained by the fact that when the ring of electrons of this material are energized they quite easily exceed the speed of light and the material goes into a higher energy state or what some people call a higher dimension. It is still there but in a higher energy state and time. Time gets all screwed up when something exceeds the speed of light. I'll tell you more about this later.

One term for these materials is micro-clusters. Micro-clusters have been described as follows on a Micro-cluster forum: "A micro-cluster is a small chemically inert cluster of atoms that has definite crystalline structure. The can be synthetic, however for this work an assumption is that the natural micro-clusters are forms comparable to the man-made micro-cluster. Micro-cluster research started with natural occurrences. Clusters exist as molecular species which can substitute and mimic various elements for one another. Micro-clusters can be as large as 200 or more atoms. Certain atomic examples for each atom are rare; the rarity is due to current physical chemistry concepts.

Research has demonstrated that natural micro-clusters are super-conductors; they are Cheshire in that they can disappear and avoid chemical detection by conventional means. Most, if not all, have catalytic properties; they are magnetic or can be induced to have electromagnetic properties; and the can from giant inert ions which I call Mega-ions."

ORMUS AND BECs

Physicists have recently created a new state of matter (which we believe is related to ORMUS0 in the laboratory. This state of matter is called a Bose-Einstein Condensate (BEC) after Satyendra Nath Bose and Albert Einstein who postulated the existence of the state of matter in the 1920s.

Their theory was not "proven" until BECs were created in the laboratory in 1995 by Eric Cornell and Carl Wieman in Boulder, Colorado. They did it by cooling atoms to a much lower temperature than had been previously achieved. This temperature was a millionth of a degree above absolute zero.

Absolute zero is the temperature at which all atomic movement ceases. When atoms are cooled near absolute zero, they move much more slowly that when they are normal temperatures. David Hudson postulates that his ORME atoms have a natural internal temperature which is very close to absolute zero. This may be why they can be Bose-Einstein condensates at room temperature and higher. A Bose-Einstein Condensate is a group of atoms which are all in the same quantum state. Such a group of atoms consequently behaves, in some ways, as a single atom. Superconsuctors are a form of BECs and so are super-fluids. Transmutation is done with frequency. The Alchemist throws in a pinch of the philosopher's stone into the melting pot and using his brain waves he starts an electron cascade. This lines up all the atoms into the same desired frequency and pouf he has gold!

You can read a simple description of what BECs are and how they work on the BEC homepage.

Here is an explanation of how BECs, superconductors and Cooper pairing inter-relate from the American Institute of Physics web page titled "BECs, Superconductors, and Cooper pairing": "A superfluid is a liquid that flows without viscosity or inner friction. For a liquid to become superfluid, the atoms or molecules making up the liquid must be cooled of "condensed" to the point at which they all occupy the same quantum state. A liquid of Helium-3, an atom whose nucleus is made up of an odd number of particles, is a type of particle known as a fermion. Groups of fermions are not allowed to occupy the same quantum state."

"By cooling the liquid to a low enough temperature, helium-3 atoms can pair up. The number of particles in each nucleus adds up to an even number, making it a type of particle known as a boson. Groups of bosons can fall into the same quantum state, and therefore super-fluidity can be achieved. Helium-4 (middle panel), a boson, does not need to pair up to for a superfluid; groups of helium-4 atoms condensed into the superfluid state at about 2 degrees above absolute zero. Superfluidity, especially the kind that exists in helium-3, is analogous to conventional low-temperature superconductivity, in which electrons flow through certain metals and alloys without resistance. In a superconductor (right panel), electrons, which are fermions, pair up in the metal crystal to form "Cooper pairs," bosons which can then condense into a superconducting state."

THE DIATOMIC NATURE OF SOME M-STATE MATERIALS

The following elements, which are known to have an m-state, have an odd number of electrons and protons Cobalt, copper, Rhodium, silver, Iridium, and Gold.

In order for these atoms to be superconductors in the m-state, they must be at least diatoms.

The m-state of gold and other precious elements is different from the metallic state of these same elements. For example, ingesting m-state gold has different effects on the body than the effects of ingesting metallic gold. What makes the ORMUS state atoms different is that they will not for metal-metal bonds with their own kind.

They won't form metal-metal bonds because their valence electrons are not available to form normal molecular bonds. This is because each electron is paired up with another electron in a Cooper paired state. When electrons are Cooper paired, the cease to behave as particles ad begin to behave more like light. Since you have an even number of electrons in order for every electron to pair up with another electron, you cannot have the m-state of any element which has an odd number of electrons without having at least tow of the atoms paired up.

For example, iridium has an atomic number of 77. This means that iridium has 77 electrons. 76 of these electrons could pair up but that would still leave on electron available for bonding with another atom in a compound. But if you had two atoms of iridium with mingled nuclei and electron clouds you would have 154 electrons. Since 154 is an even number, all these electrons can pair up into 77 Cooper pairs. Nucleons also pair up in the same way to form superconductors. All known superconductors involve this kind of Cooper pairing.

Please realize that as a Bose-Einstein condensate, both atoms in the diatom will behave as one atom. They also resonance couple with other diatoms of the same element which are nearby. The resonance-coupled quantum oscillation is another of the definitions of superconductivity. As you use chemistry to move a metal toward the ORMUS/BEC state, the chemical reactions necessary to do this moving become weaker and weaker since fewer and fewer of the valence electrons are available to participate in the chemical reactions. Eventually there are no electron handles that can be used to manipulate these materials. Fortunately these material have other properties which can be used to manipulate them.

Since they are superconductors, they can be manipulated by magnetic fields. For example, if you shield them from magnetic fields during boiling process, you will be able to conserve more of the in you liquid since they will no be impelled to tunnel out of your container or go off as a gas. They can also be manipulated by providing them with a comfy "box" to hide out in. The ORMUS/BECs seem to "like" tight spaces. Ring molecules such as the tri-sodium ring or the diozone ring can provide a chemical "box" with handles. Salt and sodium, in particular, seem to stabilize the ORMUS materials, theoretically by forming a triangular structure or box around the precious element atom. Though you cannot get a chemical handle on the fully Cooper paired ORMUS atoms, you can entice them into a chemical box with handles and then manipulate the box using fairly standard physical and chemical methods.

So, although these elements are the same as the "heavy metal" elements, they are not in a metallic state and as long as the m-state of these elements is present in sufficient amounts, the metallic portion seems to "borrow" the properties of the m-state.

BECs are also known to have the ability to "tunnel" across impenetrable barriers. Professor Brian D. Josephson of the Theory of Condensed matter Group of the Cavendish laboratory Cambridge (i.e. the Physics Department of the University of Cambridge) received the Nobel Prize in physics for his discovery of the tunneling phenomenon. Dr. Josephson is currently working on something called the Mind-Matter Unification Project.

ORMUS AND MICROTUBULES

Other physics are also working on theories which unite mind and matter. One fairly recent discovery in biology and physics is that a certain small structure in every cell, called the microtubule, exhibits superconductive and tunneling behaviors at body temperature.

You can read more about the quantum properties of microtubules from links on Rhett Savage's Quantum Brainweb-page. Matti Pitkanen's web pages:

Exotic atoms and a mechanism for superconductivity in bio-systems. Nebentropy Maximization Principle and TGD inspires Theory of Consciousness.

TGD: Eish model for the EEG and generation of nerve pulse.

One of the problems with modern quantum physical theories is that there is no way to logically connect the Bose-Einstein condensates, which have been demonstrated to exist in small groups of atoms at a millionth of a degree above absolute zero, with the BEC like behavior of microtubles at body temperature in living in cells. ORMUS material would make this connection.

Several of the modern theories relating to microtubules were proposed by Roger Hameroff's theory quite elegantly:

"Penrose has been seeking a better way to explain the fantastic computational power of the brain and Hameroff had been seeking the source of consciousness. The two heard of each other and got together to find that they both sought a common structure, the microtubule.

"Penrose sought a structure in the brain that had nanometer dimensions because such a structure would be necessary to support quantum effects. Hammeroff sought the structure responsible for consciousness. They agreed that the microtubules would provide for both.

"Microtubules are tiny tubular structures within neurons that are made from two forms tublin. The two forms can be switched by tiny currents, so Penrose has proposed that the tublin units may be the on'off switches for the brains data processing. I agree with this proposal because it allows up to be what we are by increasing our potential processing rate from an unacceptable 10 Exp11 operations per second (OPS) to a reasonably acceptable 1 Exp24 OPS. Penrose explains all this quite well and I recommend him to all who would like to have a deeper understanding of our minds.

"Hameroff has done a lot of research into how consciousness works and he has concluded that the microtubules are the souces of our consciousness. This is discussed in and supported by Penrose's work. Hammeroff has concluded that the observable quantum effects that occur in human brains are caused by highly aligned water that is inside the microtublules. Penrose agrees with concept and further argues that Bose-Einstein condensations (BECs) in the neurons are how we reach decisions. The BECs are possible because the water inside the microtubules can be strongly aligned to form a high-temperature superconductive medium.

"This concept supports my thinking very well. BECs provide an explanation for all the effects I refer to as psionics. These effects include: telepathy,remote viewing, biolocation, telekinesis, and astral travel. A BEC in the Broca area of the brain would allow thoughts to exist inside the brain and outside the head at the same time. This can explain telepathy and how it is controlled.

Likewise, a BEC in the visual processing areas would explain reomote viewing. Since microtubules exist in all neurons and extend into all parts of the body, a BEC including all neurons would allow the body to exist in two (or more) places at the same time, thus explaining bio- location.

"With the discovery, all psionics can be explained in modern physical terms. This opens the whole field of psionics to persons like myself who have had so much technical training that it is impossible to accept psionics. This discovery means that all the formal training that I've had in Chemistry, Mth and Physics still applies and can even help explain psionics. For me, it is food to know that all these topics can exist peacefully together."

For more information on "psychic" observations of these materials see: Paranormal Observations of ORMEs Atomic Structure (in the White Gold web site)

In a paper titled "Orchestrated reduction of quantum soherence in brain microtubules: A model for consciousness" Hameroff and Penrose write:

A critical number of tubulins maintaining coherence within [microtubules] for 500 m-sec collapses its own wave function (objective reduction: OR). This occurs because the mass-energy difference among the superpositioned states of coherence tublins critically perturbs space-time geometry. To prevent multiple universes, the system must reduce to a single space-time by choosing eigenstates." Hameroff and Penrose are saying that in order to avoid "seeing" multiple universes at the same time, the quantum coherence created in microtubules by some material (we think the M-state materials) must collapse. What if the quantum coherence did not collapse and we became aware of multiple universes?

Many modern physicists believe that there re an infinite number of parallel universes. They theorize that atoms are made up of smaller particles which are like bubbles in the quantum foam. These bubbles in the quantum foam or "holes" in the aether" spend a fraction of their existence in each of these parallel universes. You can read a bit about this debate at: Braintennis http://www.hotwired.com/synapse/braintennis/97/41/indesOa.html

This concept of multiple or parallel universes has been a recurring theme in science fiction for at least 60 years. It is also on of he key concepts of modern mystical thought. It first appeared as a mystical concept in the 'Unknown' Reality by Jane Roberts which was dictated by Seth in 1974-75.

SOME TESTS FOR M-STATE

M-state material, in a wet precipitate form, will dissolve in HCL. M-state material, in a dry powder form, will not dissolve in HCL or aqua regia.

Because M-state is a superconductor, rotation a magnet under some of the dry powder M-state elements will cause the powder to fly away from the magnet.

INFORMATION RESOURCES

For the interested person willing to do the necessary work, there is an abundance of detailed Technical information available on the Web. The best related web sites are: High-Spin Monatomic
Research at: http://monatomic.earth.com/

The White Gold Web Home Page at:
http://www.zz.com/whiteGoldWeb/

Superconductive s-ORMEs
http://www.jps.net/shacoma/dh/

ORMUS CHEMICAL PRODUCTION TECHNIQUES WARNING!

This document may not be reproduces except in its entirety, and without changes. Before trying any of the procedures described in this document, we advise you to thoroughly read this document several times.

This document was crated by a group of people who believe that this information is of inestimable value to humanity and should be made available as soon as possible. The information here is declared to be in the public domain and we wish that it not become the sole property of any individual or group.

Here we describe some simple ways of making ORMUS so that readers can begin true scientific and intuitive experiments with these material.

All of these methods are experimental. The following information is presented to promote scientific research into the nature of these materials. Although there methods are based on our best knowledge at this time, further scientific research may prove some of these processes or theories to be in accurate.

DISCLAIMER

The processes described here have not been tested extensively. We do not guarantee the procedures in this document, nor the results obtained by using them. To the extent that you use or implement these procedures or the products thereof, you do so at your own risk. In no event will the authors or this document be liable to you, anyone else, or any organization or government, for any damages arising from your use, or your Liability to use these procedures or the product thereof. Apply these procedures a you own risk.

VERIFICATION

The material made by some of these methods has been tested by an independent lab using x-ray fluorescence and photo spectrometry to identify the emission spectra of m-state materials. (The lab prefers to remain anonymous). The m-state spectral emissions signature was a broad, flat band rather than discrete lines. The test also showed a significant amount of calcium and magnesium, but no toxins were evident in well-washed material made from unpolluted ocean water.

To further prove that these materials are a different state of the precious elements mentioned above, it is possible to electroplate these elements out as precious metals.

People familiar with Hudson's process claim that the materials produced using these methods are similar to Hudson's ORME materials.

INGESTION

We do not recommend the ingestion of these material since so little is known about them. This information is being provided so that scientific inquiry can commence into the nature of these materials. We realize that, despite recommendations to the contrary, some people will ingest these material. With this in mind we offer the following information to minimize any possible adverse effects from ingesting these materials. Please read the WARNING AND CAUTION sections.

Some people have ingested the m-state material made by these methods. They suggest that benefits are most likely when dosage is kept small.

Three methods of making ORMUS are described in this document: the WET method, the DRY method, and the BOILING GOLD method. For the materials extracted by the wet and dry procedures, one teaspoon of material, morning and evening, has been found by them to be not harmful over several weeks' time.

A much smaller dose, on the order of a few drops a day, would be more appropriate for the material produced by the boiling-gold method. We believe that the m-state may be homeopathic, so a much smaller dose may be safest-such as 1/64 teaspoon diluted in one quart of pure water, taken two or three ounces once or twice a day.

David Hudson gave some information on dosage in his lectures.

Dallas speech at:
http://monatomic.earth.com/david-hudson/1995-02dallas-toe.html

WHITE GOLD WEB PAGE

You can find a discussion forum on the White Gold Web page. There you can post com ments and questions on these procedures and on ORMUS in general: http://www.zz.com/WhiteGoldWeb/

INDEX

1. Overview
2. Necessary Supplies
3. pH paper of pH meter
4. Safety
5. Wet method and Starting Materials.
6. Problems Encountered
7. Avoiding problems.

OVERVIEW

This document describes three methods of producing ORMUS: The WET method, the DRY method, and the BOILING

GOLD METHOD

All three methods use a chemical lab technique called "measuring pH." The pH of a solution is a measure of its acid/base ratio. You may remember testing pH with litmus paper in high school. pH values less than pH 7 indicate an acid. Like distilled white vinegar. pH 7 is neutral, like pure water. Greater than pH 7 is alkaline, like lye.

ORMUS precipitates between pH 8.5 and 10.78.

The WET method produces the last "effective" material but is relatively simple to perform.

HERE IS THE BASIC WET METHOD IN BRIEF. IT WILL BE *DISCUSSED LATER IN DETAIL:*

1. Start with drinkable water or clean sea water.
2. Slowly ass a solution of lye mixed with water to raise the pH above 8.5 but no higher than 10.78.
3. A white fluffy precipitate will form which you should allow to settle overnight.
4. Remove the liquid above the precipitate.
5. Thoroughly wash the precipitate. It is calcium, magnesium hydroxide, and a small amount of m-state material.

HERE IS THE DRY METHOD IN BRIEF:

1. Start with dry mineral powder.
2. Boil it in lye water at pH 12.
3. Filter and discard the precipitate.
4. Add distilled white vinegar or hydrochloric acid (HCL) to the filtered liquid to lower the pH to 8.5.
5. Let the precipitate settle overnight.
6. Remove the liquid above the precipitate.
7. Wash the precipitate. That is calcium hydroxide, magnesium hydroxide, and a small amount of m-state material.

AND HERE IS THE BOILING GOLD METHOD IN BRIEF:

1. Boil gold dust in lye solution.
2. Filter out any solids.3. Add distilled white vinegar or HCL to the remaining liquid to lower the pH to 8.5.

3. Let the precipitate settle overnight.
4. Remove the liquid above the precipitate.
5. Wash the precipitate. It is almost pure gold m-state material.

NECESSARY SUPPLIES TO MAKE M-STATE.

USE A GLASS OR STAINLESS STEEL POT.

If you use stainless-steel pots, check for steel particles in your precipitate. Although unlikely, this problem may occur if you use large amounts of HCL to lower the pH. Never use aluminum containers or utensils because aluminum will react with acids like HCL and alkalis like lye, and will poison you.

Distilled water from a grocery store. A stainless steel spatula or knife for stirring, from a grocery store. Never us aluminum!

Lye (sodium hydroxide or NaOH) We will use the term "lye" in this document rather than "sodium hydroxide" or "NaOH' sin it is shorter and more familiar to most people. Grocery store lye, such as Lewis Red Devil Lye, is not as pure and uncontaminated as laboratory or food-grade lye. We strongly recommend that laboratory or food grade-sodium hydroxide be used if the m-state is intended for ingestion since grocery store lye may contain dangerous contaminants. Note: Virtually no lye will be present in the final product so it will be safe to ingest. In any case lye is not toxic, and it is not caustic when sufficiently diluted (as in these methods).

HCL (hydrochloric acid or muriatic acid). We will use the term "HCL" in this document rather than "hydrochloric acid" or "Muriatic acid" since it is shorter. You can use muriatic acid (31% HCL) from a hardware store, but laboratory, electronic or good-grade HCL is less likely to be contaminated. We strongly recommend that laboratory, electronic or food-grade hydrochloric acid be used it the m-state is intended for ingestion since muriatic acid from a hardware store may contain dangerous contaminants. The presence of iron as a contaminant in the acid may interfere with the m-state materials in some applications.

Get three eyedropper bottles from a pharmacy. An alternative to eyedroppers is squirt bottles made of HDPE. Find them at a natural foods store or other store which sells bulk liquid products like vegetable oils or lotions.

Use a large 50 cc plastic syringe from a veterinary supply shop or a lab-supply house. Some supplies are listed near the end of this document under LAB SUPPLIES. pH paper or a pH meter. You can get pH paper (pH 1 to 12) from a lab-supply company or a mining supply store. Use new paper because old paper becomes in accurate. Some supplies are listed near the end of this document under LAB SUPPLIES. pH PAPER OR pH METER?

Some experimenters say not to rely on a pH meter because its readings vary with temperature and ionization.

Also, a meter costs much more than pH paper. Many pH meter probes can be damaged by very strong acids or alkalis. But some say that a pH meter is essential, for thesereasons: pH paper cannot track rapid changes in pH.

pH paper does not resolve pH readings finely enough. It's hard to tell the difference between pH 9.5, 10.0 and 11.5. pH meters are best used to get accurate readings between pH 8.5 and 10.78, which is the main range of concern in these methods. pH meters can spot check any reading with a standard buffer solution.

A pH meter is more convenient. Use only a meter that has an automatic temperature-correcting function up to 100 degrees C.

SAFETY TIPS

Clean your containers so that you'd feel safe drinking our of them. Boil containers, syringes, siphons and so on before use to sterilize them.

CAUTION!!

Lye can damage the eyes by rendering the cornea opaque, a form of eye damage that is irreparable. lye can burn skin, clothes and eyes. Work near a sink, facet, or other source of wash water. You might keep a spray bottle of distilled white vinegar hands to use against spills.

If you spill lye on your cloths or body, immediately wash it off with lots of water. When working with lye, avoid touching your face or rubbing your eyes. Do not handle lye around food. Use Adequate ventilation such as a range hood. Do not dump waste water on the ground. Lye is generally safe to put down the drain, but don't mix it with acid that may be in the drain as it can react explosively. When working with lye, please wear goggles or a full-face visor. (Industrial face protector), neoprene gloves, and a PVC lab apron. Sources for this safety clothing are in the Appendix near the end of this document.

Keep children and pets away from the work area, and do not leave it unattended if children or pets are around.

Glass can shatter with hot liquids. Pour boiling liquid from your heating container into a stainless steel mixing bowl to cool before pouring the liquid into a glass container.

THE WET METHOD - STARTING MATERIAL FOR THE WET *METHOD*.

Some starting materials produce a lot of precipitate, while other do not. Listed below are materials that have been shown to produce some precipitate from the WET method.

Some municipal drinking water.

Some hot springs water without sulfur.

Trace Minerals Inland Sea Water.

Urine

Some lake or river water whose bed or course is limestone.

Some well water. Ground water is probably more likely to contain m-state than surface water (except seawater).

Seawater and sea water reconstituted from certain brands of sea salt, especially from The Great Salt Lake.

Dead Sea water.

Certain brands of unrefined sea salt are as good as sea water: Celtic Gray Sea Salt (from health Food stores) and Lima Atlantic Sea Salt (from some health food stores). Add distilled water and use the WET method. Filter the scum first.

The WET method performed on ocean or dead Sea water produces eleven different m-state elements. The following materials are ranked in order from most to least m-statecontent:

1. Dead Sea water.
2. Salt Lake water.
3. Ocean water.
4. Well water.

Listed below are material that have been found to produce little or no precipitate from the WET method:

Water from some alkali lakes (pH above 8.5).

Hot springs with sulfur (because sulfur reduces m-state to metal). Mineral-free lake or river water.

Dead Sea mineral salts that contain sulfur or sulfates, such as "Sea Mineral Bath from the Dead Sea" by Dead Sea Works Ltd. For Sea Minerals Co., and Trace Minerals Research "Concentrate Trace Mineral Drops" from the Great Salt Lake.

For the following methods to work, some researchersclaim that magnesium or magnesium hydroxide - Mg (OH)2

- • must be present in the starting material. (Since the Boiling Gold Method is effective without and magnesium, this claim will need to be tested). Sea Water already has Mg(OH)2, so you don't need to add it to sea water. Try your water first. If you don't get any precipitate, you might add a teaspoon per gallon of Epsom salts to the starting material for its magnesium. If you do add Epsom salts, the magnesium from them will be a large portion of the precipitate.

WARNING! - PROBLEMS ENCOUNTERED

The following problems have been encountered by some folks who have made m-state for consumption.

Some people have gotten quite sick from consuming m-state made from sea water collected at a marina.

This water contained high levels of lead and other contaminants. Other people have gotten sick from consuming m-state which are made improperly. These material were made without the use of pH test paper or meters and the resulting material contained toxic metals. Please remember that old pH paper can become inaccurate.

People have gotten sick from consuming m-state materials which contained bacteria because they were not sterilized or stored properly.

It is possible to bring the pH of your source material up too quickly, especially is you use lye in too high a concentration. This could result in local areas of very high pH within your solution. These high pH areas could allow toxic metals to precipitate and mix with your desired precipitate.

M-state platinum might be considered toxic by some since it make you quite ill if you consume alcohol. No one has reported this effect from consuming m-state from sea water.

Some people have used Teflon* (Registered Trademark) coated aluminum sauce pans for heating lye or lye water. The Teflon* (Registered Trademark) got scratched and the aluminum started dissolving in the lye water producing hydrogen gas which could have exploded. The liquid was contaminated with aluminum which is a poison.

AVOIDING PROBLEMS

Use sea water, reconstituted sea water made from sea salt or Dead Sea salt, or lake water. In general, start with a clean and deep source of water. Some people have gone out to sea in boats to collect sea water from 100 feet deep.

Generally avoid water that has lead, arsenic or other toxic elements in it. Start with water that is drinkable except for salt content.

Conduct and elemental and toxic analysis of questionable starting-material sources (such as sea water collected close to the shore, or near sources of industrial waste runoff).

Boiling in lye water kills bacteria but it does not destroy toxic metals or chemicals in your source water.

Follow these instructions and slowly change the pH of your solution.

Avoid water with sulfur or sulfated in it because such water produces little or no m-state precipitate.

Never use aluminum containers or utensils because aluminum will react with acids like HCL and alkalis like lye, and could poison you.

WET-METHOD PROCEDURE

Please read CAUTION!! and WARNING!! Before proceeding First you need to prepare a dilute solution. Label and eyedropper bottle or squirt bottle (Lye-poison" so the bottle will not be confused with something else. Work in a sink so that any spills will be contained. Lye gives off eye-stinging fumes when mixed with water. To avoid inhaling fumes, hold your breath and wear goggles while doing the following procedure.

Working over a sink, put 8 teaspoons of distilled water in a sturdy glass then stir in 1 teaspoon of lye. Stir until the lye is dissolved. Heat will be generated as the lye dissolves and the glass may get fairly hot. You may want to close your eyes to avoid eye-stinging fumes, taking a peek periodically. Pour the lye solution into a labeled eyedropper bottle or squirt bottle.

If you are using pH paper, tear off several ¼" pieces and put them on a piece of white paper on a plate.

For the best accuracy, recalibrate the pH paperthroughout the day with changed in temperature and humidity, as well as day-to-day. Buffer solutions of pH 4,7 and 10 will help with this. Sources of pH buffer solutions are listed near the end of this document under LAB SUPPLIES. If you are using dried mineral, mix ½ cup of dry material with 2 cups of distilled water. This makes sea water. Now proceed as described below.

1. **First, you might want to pour the sea water through a coffee filter to remove any scum.**

2. **If the starting material does not contain magnesium** hydroxide (sea water does contain magnesium hydroxide), add some, or add a teaspoon of Epsom salts ore gallon of water.

3. Pour the sea water into a stainless steel pot. Slowly, drop-by-drop, ass the lye solution WHILE STIRRING.

    Every ten drops or so, test the pH. You might want to take at least 3 to 5 samples from different regions of the liquid. If you are using pH paper, the goal is to bring the pH up to p.5, then stop to be on the safe side. If you are using a pH meter, stop just before you get to pH 10.78.

4. Once you are at the correct pH, stop.

5. Pour the solution into a clean glass jar or test tube.

6. The white precipitate (slurry) slowly settles on the bottom of the jar. Let the slurry settle over night. If metals or other toxins have been ruled out by prior testing of your starting material, the slurry is probably mostly calcium hydroxide, $Mg(OH)2$, lye, and a small amount of m-state.

7. You can speed this settling process with a centrifuge, which forces the precipitate to settle rapidly.

8. **Inexpensive second-hand centrifuges may be found at**

    American Science and surplus.

9. http://www.sciplus.com

10. Using a large syringe (or siphon), remove the liquid above the slurry.

11. Ass distilled water to the precipitate (filling the jar), stir thoroughly, and let it settle again for at least 4 to 5 hours, preferably overnight.

12. Repeat steps 7 and 8 at least three times to thoroughly wash the precipitate. This should remove almost all of the lye. The remaining lye can be neutralized with HCL or distilled white vinegar as well. Washing three times is intended to reduce the dissolved "impurities" (like salt, for example) by 87.5%.

Four washes would provide 93.75% reduction, five washes a 96.87% reduction, and so on.

At this point, the precipitate is likely to contain some, m-state, mile of magnesia Mg(OH)2, calcium, and perhaps some impurities.

Pour the precipitate and water into a stainless steel pot on a stove burner. A gas burner is preferred over electric because any magnetic fields from the electric burner may drive off some of the m-state material. Cover the pot with a lid to contain the m-state, and boil the solution for 5 minutes to sterilize it. Be careful not to spill the hot solution! Let it cool back to room temperature and recheck the pH to make sure it hasn't exceeded pH 9.

DISCUSSION: WHEN TO BOIL THE SOLUTION

In this document, we suggest that you not boil the solution until you have made the washed precipitate. However, boiling can be done earlier in the procedure with certain advantages. Here are four times that boiling can be done, with a discussion of the pros and cons of each:

1. **Boil before adding lye solution. PROS: Faster reaction,** faster precipitation. CONS: You may spill the hot lye solution. You may inhale fumes: Danger of lye spurting our of pot. Not recommended.

2. Boil while adding lye solution. PROS: Faster reaction, faster precipitation. CONS: You may spill the hot lye solution. You may inhale fumes. Danger of lye spurting out of pot. Not recommended.

3. Boil and add lye solution. PROS: No danger of inhaling fumes. Little danger of spilling hot lye solution. CONS: Slower reaction, slower precipitation.

4. Boil the washed precipitate (recommended). PROS: No danger of inhaling fumes. N danger of spilling hot lye solution, pH is unlikely to change after boiling because the reaction has already taken place. CONS: Slower reaction, slower precipitation. If safety is the main issue, this seems to be the best method.

CAUTION: If you boil the solution on an electric burner, the magnetic field in the burner may "blow off" some of the m-state materials, resulting in a small yield. This can be minimized by adding a source of sodium (such as sodium hydroxide or salt) to the solution before boiling.

Since sea water contains sodium in salt, none of the boiling methods will be a problem with sea water. However, if you are starting with low-sodium fresh water, ass a sodium source (such as table salt or lye solution) before boiling.

Once the precipitate and water have been sterilized, the next step is required to concentrate the m-state.

HOW TO PURIFY YOUR PRECIPITATE

The precipitate made from sea water contains milk of magnesia ($Mg(OH)_2$), which precipitates approximately around the same pH range that m-state does.

Here are four methods to separate $Mg(OH)_2$.

METHOD 1

1. Suppose you just made a precipitate by adding lye solution to sea water. The precipitate is m-state mixed with $Mg(OH)_2$.

2. Use a syringe to remove the liquid over the precipitate, and discard the liquid. This leaves only the m-state/$Mg(OH)_2$ precipitate.

3. To the wet precipitate, add hydrochloric acid (HCL) until you reduce the pH to 1.0 - 3.5. You can use muriatic acid (31% HCL) from a hardware store, but lab grade HCL is less likely to be contaminated. A safe alternative to HCL is distilled white vinegar.

4. The white colloidal precipitate should dissolve, leaving a clear solution.

5. Add lye solution VERY SLOWLY drop-by-drop to bring the pH back up to 8.5 - 8.7. The precipitate that forms should be m-state mostly free of $Mg(OH)2$ (because m-state precipitates in this pH range, and $Mg(OH)2$ does not precipitate until pH 9. Note that your total yield may be diminished because you are not going past pH 8.7.

6. Remove the liquid above the precipitate, and wash the precipitate. It should be mostly m-state.

METHOD 2

This procedure removes the $Mg(OH)2$ by dissolving it below pH 9. First get some HCL (or muriatic acid) and coffee filters. A safer alternative to HCL is distilled white vinegar.

1. Dry the precipitate in a dark oven at about 275 degrees F for one or two hours. This forms a dry powder.

2. Take the dry powder and pulverize any clumps.

3. In a glass container, cover the powder with some distilled water. For example, one liter of water for one cup of powder.

4. Add HCL or distilled white vinegar drop-by-drop to bring the pH to 5 or 6.

5. Shake the bottle and let it sit overnight. The dried m-state should not dissolve at that pH, but the $Mg(OH)2$ should dissolve.

6. The next day, after all the Mg(OH)2 has dissolved, pour everything into filter paper.

7. Wash the powder collected in the filter paper several times with distilled water to remove any residual traces of HCL or vinegar.

8. The washed powder may be oven-dried again about 275 degrees F, and you should have m-state powder free of Mg(OH)2.

METHOD 3

1. Dry the original precipitate at about 200 degrees F.

2. Mix the resulting powder with distilled white vinegar or 30% HCL. Everything which does not dissolve is m-state. This will be quite a small amount if you start with sea water. (If you mix pure HCL with distilled water, remember: ADD ACID TO WATER, NEVER ADD ***WATER TO ACID.).***

3. Measure the amount of HCL/m-state solution (or vinegar/m-state solution).

4. Add distilled water to the HC/m-state solution. Add an amount of water that is at least ten times the amount of HCL/m-state solution. (You may substitute distilled white vinegar for HCL).

5. Filter the solution through 5 layers of coffee filters.

6. Wash the powder at least three times in a large amount of distilled water.

METHOD 4

1. Starting with clean wet precipitate, add lye to bring the pH up to 12. The m-state precipitate will dissolve, but magnesium hydroxide and the Gilcrest precipitate will not.

2. Filter out the precipitate.

3. To the remaining liquid containing only m-state, add HCL or distilled white vinegar drop-by-drop until the pH reaches 8.5.

4. Add lye solution drop-by-drop to bring the pH back up to 10.78. The resulting precipitate should be only m-state.

5. Wash the precipitate as described earlier.

6. To be safe, check the pH of the precipitate slurry. It should be 9 or less before ingesting.

DRY METHOD

Please read CAUTION!! And WARNING!! Before proceeding This method takes longer than the WET method. In some cases, it involves boiling lye for several hours, which may spray some caustic solution around your work area. Please wear neoprene gloves, a PVC lab apron, and eye goggles when you use this method. Sources for this safety clothing are listed near the end of this document under LAB SUPPLIES. Some people have reported adverse reactions to the WET method precipitate or powder. This may be due to the Gilcrest precipitates, so it results in safer material.

EXTRA SUPPLIES NEEDED FOR THE DRY METHOD

12-cup coffee filters from a grocery store.

Hydrochloric acid. You can use muriatic acid (31% HCL) from a hardware store, but lab-grade HCL is less likely to be contaminated. Other acids can be used, but HCL will not harm the body if accidentally ingested in weak solutions and in small amounts. You might prefer to use distilled white vinegar instead of HCl. Although distilled white vinegar (acetic acid) is weaker than HCL, and is safer to work with. Heavy plastic HDPE cottage cheese containers, 1 pint and 1 quart, to hold the coffee filters.

MAKING A HOLDER FOR THE COFFEE CONTAINERS

1. Start with a pint and a quart container for cottage cheese. Make sure the pint container will fit the quart container. The pint container should hang inside the lip of the quart container.

2. Across the bottom of the pint container, punch or drill several holes, 1/8" to ¼" diameter, about ¼" apart.

3. If the small container fits too tightly into the larger container, you may need to drill some air-pressure equalization holes around the outside of the larger container near the level of the bottom of the small container. Otherwise the air pressure between the two containers will keep liquid from draining from the coffee filters. When you use this filter, place the cottage cheese containers in a stainless steel or glass container to catch any overflow. The lye water that you will be filtering may damage counter tops or cabinets if it contacts them.

4. The coffee filters should fit nicely into the smaller top cottage-cheese container.

DRY METHOD STARTING MATERIALS

Generally start with dry material such as sweepings from salt and alkali flats, rock powders, limestone, mineral salts,

Isis or Etherium white gold powder, volcanic ash, plant cinders, etc.

These are some material that produce a lot of precipitate from the DRY method: Crushed, unheated limestone (CAUTION: agriculture grade powdered limestone from some sources contains sufficient lead and / or arsenic to be potential hazard) Golden Nectar trace mineral formula. Etherium/Isis Gold powder

Ancient Secrets Dead Sea Mineral SaltsMasada salts (unscented).

DRY-METHOD PROCEDURE

Please read CAUTION!! And WARNING!! Before proceeding First you need to prepare a diluted lye solution. Label an eyedropper bottle or squirt bottle "Lye-poison" so the bottle will not be confused with something else. Work in a sink so that any spills will be contained. Lye gives off eye-stinging fumes when mixed with water. To avoid inhaling fumes, hold your breath and wear goggles while doing the following procedure.

Working over a sink, put 8 teaspoons of distilled water in a sturdy glass then stir in 1 teaspoon of lye. Stir until the lye is dissolved. Heat will be generated as the lye dissolves and the glass may get fairly hot. You may want to close you eyes to avoid eye-stinging fumes, taking a peek periodically. Pour the lye solution into a labeled eyedropper bottle or squirt bottle.

If you are using pH paper, tear off several ¼" pieces and put them on a piece of paper on a plate. Now proceed as described below.

1. Grind the starting material to a fine powder.

2. Add 1:4 lye solution to cover the dry material with a thin layer.

3. Stir in some distilled water to cover the powder and lye by 2 inches.

4. Bring to a boil (this is best done outdoors or in an exhaust hood). The pH should be at or slightly above 12. The lye brings the m-state elements into solution while leaving the Gilcrest precipitates as solids.

NOTE: If you start with sea salt, you can omit the boiling step with its noxious fumes, and simply let the solution sit for three days. Then go directly to Step 7. (Some other starting material might also react without boiling).

1. If you are boiling the solution, replace water as needed to maintain sufficient reactant volume.

2. Boil for several hours-the longer the better-in a closed container. This container may be open if you add liquid needed. Four hours should be sufficient for Etherium/ Isis material.

3. Strain the slurry through 3 to 5 layers of coffee filters. You are removing the toxic elements (Gilcrest precipitate) that precipitates above pH 11.5. Save the liquid that passes through the filters. Most m-state present will be in solution in the liquid.

4. While stirring the liquid, slowly add HCl or distilled white vinegar to bring the pH down to 8.5. A white precipitate forms which is

partly m-state. If you go too far, the pH will abruptly shift, and you will have to start over. If this happens, you must add lye quickly and bring the pH back up to 12.

5. Let the precipitate settle overnight.

6. Using a large syringe (or siphon), remove the liquid above the slurry.

7. **Add distilled water to the precipitate (filling the jar),** stir thoroughly, and let it settle again for at least 4 to 5 hours, preferably overnight.

8. Repeat steps 10 to 11 at least three times to thoroughly wash the precipitate. This removes most traces of lye and HCL (or vinegar).

You'll get a wet, white precipitate (slurry) containing m-state elements. Check that the pH is 9 or less before ingesting. Some of the precipitate may be milk of magnesia or calcium. If you wish, you can remove them using the precipitate purification procedures described above.

BOILING GOLD METHOD

Please read CAUTION!! and WARNING!! Before proceeding This method produces pure gold ORMUS. With this method, you must boil some lye solution for one to two weeks. This may spray some caustic solution around your working area. Please wear neoprene gloves, a PVC lab apron, and eye goggles when you use this method. Sources for this safety clothing are noted under LAB SUPPLIES near the end of this document.

EXTRA SUPPLIES NEEDED FOR THE BOILING-GOLD METHOD

Coffee filters from a grocery store.

Hydrochloric acid (HCL). You can use muriatic acid (31% HCL) from a hardware store, but lab-grade HCL is less likely to be contaminated. Instead of HCL, you might prefer to use distilled white vinegar (acetic acid). Distilled white vinegar is weaker than HCL but is safer to use.

A stainless steel pot will work, but stainless steel and glass are attacked by NaOH. Glass is preferred over stainless steel.

A preferred container is a sealed Teflon® or HDPE bottle in a water bath in a crock pot. Please note that a Teflon® coated bottle is NOT the same as a Teflon® coated aluminum container. Never use aluminum or Teflon® coated aluminum containers and utensils because aluminum will react with acids like HCL and alkalis like lye, will poison you. If you use a sealed Teflon® pr HDPE bottle in a water bath fill it only half full with lye and gold then squeeze the bottle to eliminate most of the air above the liquid before you tighten the cap. This will allow the bottle to expand as the liquid is heated.

PROCEDURE FOR THE BOILING-GOLD MTEHOD

1. Add 99.99% pure gold dust to a lye solution of pH 12 or more.

2. Boil the solution for two weeks in a CLOSED container. One week may be sufficient, but two weeks will likely have a higher yield. Add water as needed. CAUTION:

    Do not inhale the vapors.

3. Strain the solution using the coffee filter holder described above. Save any remaining gold for future use.

4. Add HCl or distilled white vinegar to bring the pH down to pH 8.5. An off-white precipitate will appear. Let it settle overnight.

5. Using a syringe, carefully suck out the liquid above the precipitate.

6. Add distilled water to the precipitate (filling the jar), stir and let it settle again for at least 4 to 5 hours, preferably overnight.

7. Repeat steps 5 and 6 at least three times to thoroughly wash the precipitate.

APPENDIX: M-STATE STORAGE

Glass mason jars with wire-clamped glass lids and rubber gaskets? Glass jars with plastic lids, or HDPE containers, which are stable in acid and alkali Store m-state material in the dark away from sunlight or ultraviolet light. Ultraviolet light seems to move some -m-state material toward a metallic state.

Because m-state materials are superconductors, they should be stored in glass HDPE containers inside of steel containers and away from moving magnetic fields. Put the glass or HDPE container (containing m-state) inside a steel tin such as those used for Christmas cookies, gourmet popcorn or potato chips. If you intend to transport m-state materials, it is best to nest three or four steel containers, one inside the other with insulating material between, and place the glass or HDPE container inside the inmost steel container.

You may notice that your m-state materials have bubbles rising from them for a period of time after they are made. We believe that these bubbles are m-state gas escaping.

Some people have reported that the m-state precipitate loses some of is effect as these bubbles leave. This off-gassing seems to re reduced if the m-state is stored between room temperature and body temperature in a magnetically shielded container. It is advisable not to refrigerate m-state materials. At lease one researcher reports that refrigerated m-state materials are likely to move toward a warmer place.

Since bacteria and mold can easily grow in m-state precipitate, it is best to sterilize any material which you wish to store for long periods of time and to store it using water bath canning methods.

CHEMICAL SUPPLIERS

Mowre W. E. Co.
1242 University Ave.
St. Paul, MN 55104
800-544-1550 (646-1895)
This supplier sells gold bullion, god wire, gold shot, etc.

They sell to individuals through the mail and will process
Small orders. Good prices!
Strem Chemicals, Inc.
Dexter industrial Park
7 Mulliken Way
Newbury, MA 01950-4098
Info@strem.com
http://www.strem.com
No minimum order. (978) 462-3191 (800) 647-8736
Some sample products: 93-7915 Gold powder (99.95%) 500 mg $50. 2g
$160. 93-7913 Gold shot (99.95%) 500mg $40. 2g $128.

LAB SUPPLIES

Weiss Research Inc.
11743 West Bellfort, suite 168
Stafford, TX 77477
http://www.hia.net/weissres/meters.htm
They have a $175 pH meter (#PHM-150) with auto temperature
compensation from 0 to 100 degrees C, and a pH meter with manual
temperature conversion for $125. Mc Master-Carr Supply Co.
P.O. Box 4355
Chicago, IL 60680-4355
630-833-0300
fax 708-834-9427
Mc Master-Carr Supply Co.
P.O. Box 54960
Los Angeles, CA 90054-0960
310-692-5911
fax 310-695-2323

Plastic syringe 50cc with tapered tip 7510A665 pkg. Of 10 $18.57
Milled neoprene gloves 5307T5 $5.01
Wide-range pH test paper 8707T11 $8.89
PVC apron 53445T75 $4.55
Safety goggles with a face shield 5422T12 $18.94.

Also available are a pocket pH meter for $59, pH solutions, water-test kits, hot plates, plastic tubing, and much more. No minimum order. The sell to anyone and take VISA.

Edmund Scientific Company
101 East Gloucester Pike
Barrington, NJ 08007-1380 USA Customer service: 1-609-573-6260
9AM to 5 PM M-F
Disposable plastic syringes are available at many veterinary and agriculture supply stores. Plastic infusion tips are available from the same source.

More sources can be found at:
http://www.sigma.sial.com/

Sea salt to reconstitute sea water: Health Food store, oriental food store. Target Glacial Rock Dust from Gaia Resources in Grand Forks, BC, Canada. Get gold from a coin dealer. Get gold dust by panning for it in a stream. Gold dust might be available from Keen Engineering in Northridge, CA.

See Chemical Suppliers earlier in this Appendix. Also check out a gold mining supply store. In California try:
http://www.treasurenet.com/calgold/prospect.html

The gold should be at least 99.99% pure gold. Golden Nectar (trace minerals) cost $40/gallon.
http://bulksales.com/index.htm

This may be at health food stores. Trace Minerals Inland Sea Water ($7 for 8 ounces)at your local health food store.

Trace Minerals Research, P.O. Box 429 Roy, UT 84076
http://www.traceminerals.com
http://bulksales.com

Etherium white powder gold:
http://www.etheriumgold.com/

Phone 503-625-2880 Isis white powder gold:
http://onlinehealth.web2010.com/isis.html

SAFETY ADVICE FOR THE DRY METHOD

Target Glacial Rock Dust contains amounts of aluminum which pants can tolerate, but is toxic to people when extracted with lye. Agriculture-grade powdered limestone from some sources contains sufficient lead and /or arsenic to be a potential hazard. This form of limestone should not be used for human consumption without testing.

WARNING ABOUT GEMATRIA PRODUCTS

There seems to be some conflict between certain products of the Gematria Products company (http://www.gematria.com/) and the m-state elements. Four people who were consuming both reported symptoms including headaches, nausea, dizziness and hives which ceased when they discontinued the use of either the m-state or the Gematria products. This products which were mentioned were AloeMEM Gem™ j and Laser Blue ™.

A CAUTION RELATED TO THE WET METHOD

Some folks get diarrhea from all the m-state magnesium in the m-state produced from the Dead Sea water. The percentage of magnesium can be reduced in the final product if you pull the pH back down after taking it up to 10.78.

SOME OPTIONS FOR THE BOILING GOLD METHOD

When you put the gold and lye solution in a Teflon bottle for boiling, here are two ways to do it:

1. Fill the bottle half full, squeeze out the air, and tighten the cap. This lets the bottle expand when heated.
2. Bring the surrounding water to a boil with the bottle open. When the lye solution has started boiling, tighten the cap and continue boiling.

PERCENTAGE OF M-STATE ELEMENTS IN VARIOUS SEA WATERS

Dead Sea: gold 70%, magnesium 30%. Salt Lake: Gold 19%, Rhodium 30%, Iridium 5%, Magnesium 46%. Pacific: Gold 14%.

WHAT TO AVOID WHEN USING ORMUS

Certain things "pin" ORMUS to their metallic state: they will de-spin the monatomic Platinum-group elements into their metallic form. These should be avoided:

1. Short wavelength ultraviolet light.
2. Nitric oxide, such as in smog (not nitrous oxide).
3. Sulfides (SO3), such as in some salad dressings.
4. Carbon, such as in burned food.
5. Carbon monoxide, such as in smog. Source: Hudson interview (with Binga), June 28, 1996.
    WhiteGold@zz.com

GOOSEBERRIES ANOTHER WAY TO EXTEND LIFE

In 1970 a German Newspaper published an unusual article. This was a surprising revelation by Soviet authorities, as it was custom at this time to keep a tight lid on discoveries of this type. There was widespread speculation by western circles that this information was released by mistake.

The following is a translation of the article:
The juice of unripe Cooseberries is, according to an article in the Soviet trade-union newspaper Trud, an Elixir of Eternal Youth. The paper reports that Soviet scientists have discovered in the fruit an acid which stops the aging of cells, which is the cause of infirmities and final death of the aged, and according to Trud, even stimulates the growth of new cells.

To benefit from the above, it stands to reason that the unripe gooseberries should be ingested internally. If they are too sour for your taste you may mash them in your blender and mix with some honey or another type of sweetener. We do not recommend to juice the berries, as too much of the valuable acid will remain in the pulp.

The gooseberry goes through several unripe stages, and we recommend to harvest a cross section of unripe undersized berries. Since the gooseberry is a seasonal fruit you may want to keep a supply in the deep freezer for use during the off-season.

HOW TO GROW GOOSEBERRIES

There are two types of gooseberries: the American and the English. The latter types do nor grow well in the USA. They are very susceptible to mildew, and dislike our strong American sunshine. American varieties, on the other hand, are quite easy to grow. Poorman is a good red one usually offered by nurseries. Welcome and Pixwell are other favorites which ripen to a pinkish color. Green Mountain is a fairly reliable green or white variety. Downing, Oregon Champion, Como, Red Jacket and Carrie are among the old timers you might hear about.

Gooseberries make nice garden bushes because they won't spread all over, nor do they grow rank and tall. Sometimes, however, a few too many canes will come up, and you will have to thin them our for good air circulation. Gooseberry canes will produce fruit for 3 to 4 years before they begin to lose vigor. Usually it's best to cut them our after the 3rd bearing year. The sign of a still vigorous cane is when new growth exceeds 6 inches a year. In general try to maintain about 9 canes per bush: three of them 1 year old, 3 of them two years old and 3 of them 3 years old. Get your pruning finished very early in spring. The plants break dormancy unusually early – almost as soon as the thaw is out of the ground. For this reason some authorities advise fall planting. In the colder parts of the North, however, it may be better to wait till spring to plant gooseberries.

Gooseberries will grow in heavy as well as in sandy soils. They are not very particular about pH as long as the soil is neither extremely acid nor alkaline. Like all berries, the plants do best when mulched. Several forkfuls of manure around the base of a plant every summer is all the fertilizer it needs.

The easiest way to increase a gooseberry bush is to bend a low branch over and bury part of it in the ground, allowing the tip of the branch to stick out. The buried part will put down roots, and a year later you can cut the branch and transplant it.

Powdery mildew is probably the gooseberries most serious disease. There is not a lot you can do about it, especially organically. Keep the bushes thinned of rank growth for good air circulation, and don't plant them where they are too shaded if mildew is a problem. Since the bushes like some shade where summers are long and hot, mildew can put you in a dilemma. If it does, stick to American varieties. Anthracnose can be harmful to gooseberries too. There is no cure, but some varieties are more resistant than others. "Welcome" is the most resistant gooseberry to this blight. Various other cankers and rots may at times raise their ugly heads among your gooseberries. Usually, such diseases can be avoided simply by not fertilizing too heavily with nitrogen. Insects are rarely a critical problem on gooseberry bushes.

Sometimes gooseberries are the carrier of a disease called "white pine blister rust" which is dangerous to whit pines. But due to research and regulations the danger is by far not as serious any more as it once was. Reputable nurseries relieve you of the worry about who can and who can't grow gooseberries. But more importantly, the plants they are selling will be free of the disease.

The states without restrictions are: Alaska, Arizona, Arkansas, California, Connecticut, Florida, Idaho, Illinois, Indiana, Iowa, Kansas, Kentucky, Louisiana, Michigan, Pennsylvania, S. Carolina, Tennessee, Texas Utah and Wyoming. It is important to remember that the other states are not totally sealed off from gooseberries, only control areas within these states are. As mentioned earlier, your local nursery will be able to answer all related questions for you.

ADDENDUM

As people become smarter from the use of m-state free energy and antigravity transportation devices will become commonplace. War and differences of opinion may never be entirely resolved as long as there is a chance someone will benefit monetarily from these differences however, by expanding human consciousness there is a chance we might be able to work out some of these differences.

Many years after I wrote this book I am still discovering secrets of the MANNA. I was talking to a waitress about the manna when suddenly it hit me. The reason the million or so Hebrews could traverse the desert without starving to death was because they had the technology to make the manna. The Egyptian army could not follow them because they did not have this technology.

All plants take up energy from the soil by extruding alkaline from their roots. The alkaline dissolves the minerals and other nutrients. The reason you cannot go out behind your house and eat all the plants that are out there is because most of them have too high an alkaline content and you would be poisoned. The plants that we do eat have been genetically engineered by the Gods to have a lower alkaline content so that we can eat them.

This concept answers a few more questions such as why native plants grow better than the ones you plant. The reason is that native plants having a higher alkaline content are more efficient at taking up nutrients. The answer to the next question is, if you could eat the native plants you would be assimilating more beneficial nutrients. Do domesticated plants contain less beneficial nutrients due to their lower alkaline content? Probably so! Would we live longer if we could assimilate the nutrients directly instead of depending on plants to do it for us? Probably yes because we would be getting more of them!

Moses and his brother Aaron would boil the soil in an alkaline solution for several days. They probably used the alkaline spring waters from the spring of Murah. As the water boiled away they would have to add more alkaline spring water consequently the alkaline content grew stronger and stronger thereby making the process more effective. The liquid then became quite poisonous due to hit's high alkalinity. After pouring the clear liquid into another container they added a little vinegar to raise the acidity of the monatomic soup. This also made it safer to drink. The minerals became visible as a sort of milk. Taking this into one's body was partaking of God's light or God's food. The Hebrews probably mixed this solution with whatever grains they had brought with them and baked them into cakes to make a life-sustaining meal.

It suddenly dawned on me that this "manna food" really is the food of the Gods. A group of Angels or Gods (astronauts) traveling in space could stop at any planet or asteroid and as long as they had water they could make the manna to eat. It could even be a dead planet like Mars and they could still make the manna. This does not contradict the Bible says because the technology was given to the Hebrews by YAHWEH (Jehovah).

Obviously this technology was invented by space travelers millions perhaps billions of year ago.

Attention NASA: If you ever go on a long space voyage to a barren planet you are going to need the technology of the

MANNA

The manna made by boiling the soil in the Middle Eastern desert would contain a veritable soup of beneficial trace elements in the metallic and monatomic states. This is not the manna that Moses metal smith Bazalial made by burning gold at 5600 degrees. I think there should be more studies on the health benefits of metals and monatomic elements. We take zinc for colds and flue. We take chromium picoliniate to build mussel mass, loose weight and gain energy. A small amount of Selenium reduces the chances of a heart attack by 40%. Some of us take colloidal silver to kill bacteria and micro-virus. It also promotes new cell growth in wounds. If we ate the manna we would be getting all these things and more.

HERE IS A WAY TO MAKE M-STATE WITHOUT USING DANGEROUS CHEMICALS.

I heard of a person constructing a plywood pyramid in his back yard and hanging a gold coin in the approximate location as the King's chamber in the Great Pyramid at Giza. Underneath the gold coin he placed a glass of distilled water.

In a couple days the gold coin developed a white powder on it's surface. (Apparently m-state magnetic frequencies emanating from around the galaxy were transmuting the gold coin into m-state.)

When the coin had a good coating of m-state material he dunked it into the glass of water and drank it. Afterwards he reported that the he had taken a rather large overdose. I recommend that you drink ¼ of the water to begin with to see if you can handle the effect.

William Henry

After I wrote this book Sir Lawrence Gardner and William Henry recently made more discoveries about the Ark of the Covenant. After studying the Egyptian drawings on the walls inside the temple at Abydos they noticed a pillar or antenna-like device attached to the top of the ark. When the ORME material inside was activated, instead of the shikanna glory (ball lightening) appearing the antenna focused a beam of inter-dimensional power strait up creating a worm hole in space.

Anyone standing near the device was transported to wherever the Anunnaki had another such device set up-most likely on their home planet in the Sirius star system or possibly even the Pleiades. A picture of this device can be seen on William Henry's web site at williamhenry.net.

King Solomon had a front porch twenty stories high which housed a pillar to drill holes in space. The men who used this device were the Homo Christos or Homo Angelis. The ancient Illuminate (El-beings of light) could travel back and forth between worlds and transport whole populations of people with this simple device. EA and ENKI started the first mystery schools.

William Henry says that at the time of Jesus there were many God men or light men all over Egypt. Jesus learned many of his skills from them. He also spent many years in India learning the secrets of the mystery schools. There are many more books about Jesus in India than any place else.

Their goal was to bring forth the message encoded in your DNA.

DNA.

In the temple at Abadose the Egyptian God Set is portrayed as bringing Osirus back from the dead using a wand or rod. The wand is a branch from the tree of life (some kind of remote control device) which energizes your DNA. It is in the form of a double-spiral helix.

BIBLE HISTORY

The following is for people who would like to research our past. Whenever I write about something that touches close to a person's religions belief I wind up getting an argument about monotheism or about how the Bible is the only inspired word of God. FYI I took several college courses on Man's Religions to better understand the subject. I have read the Bible six times and written quite a bit on the subject.

It seems that many people are not aware that there are thousands of ancient books about man's past and they don't realize that the Hebrew scholars were forced to pick and choose which books should go into the Bible to suit a certain political agenda. For example, the mention of women's roll in the church was purposely left out of the Bible to keep women as ignorant sex slaves. They didn't want their women to know the history of women's role in the church.

Somewhere around 900 BC God or in this case YHWH gave Moses the Ten Commandments; only instead of ten there were actually 50. The Old Testament was originally written in ancient Hebrew.

Seven Hundred AD Benjamin Kennecott spent most of his life collected as many ancient, Old Testament manuscripts as possible eventually collecting 600.

De Rossie, professor of Oriental languages in Parma, Italy collected over 730. The manuscripts from the Cairo Geniza and the Firkowitsch collection along with the Dead Sea scrolls swelled the total collection to more than 2,200 Hebrew documents.

When Alexander The Great conquered most of the known world of his day, Greek became the Universal language. Within two or three generations many Jews had lost the knowledge of their own language. In order for them to read their Old Testament documents which were in Hebrew seventy learned Jews who knew both Hebrew and Greek, translated the old Testament into Greek. This translation called the Septuagint is often represented by the Roman numerals LXX which means "70". It was completed by at least 200 BC.

By 722 BC Samaritans (half or mixed blood Jews) were able to hang onto some ancient Hebrew manuscripts containing the first five books of the Bible known as the Pentateuch. This was before the Assyrian invasion.

Today Bible scholars call the ancient Samaritan manuscripts the Samaritan Pentateuch. There are only a few Samaritans left today. The pitiably small group still keeps some very old and valuable scrolls in the temple at Nablus, in Palestine. There are a number of excellent copies of the Samaritan Pentateuch in various libraries in this country and Europe.

The Masoretic Hebrew test, the Greek Septuagint, and the Samaritan Pentateuch are the three main channels of transmission for our Old Testament. It is very helpful to consult the Dead Sea scrolls when making a comparative study of the Masoretic Hebrew text and the Greek Septuagint.

Much like Moses who was pushing monotheism King James had the Septuagint translated into English he put himself up there with God. And, like every despot intent on fostering his their own personal views on his subjects, anyone caught with a different Bible in their possession they were killed on the spot along with their entire family.

There are at least four thousand five hundred known Greek manuscripts, ancient translations in other languages, Syrias, Coptic, and Latin versions are exceeding not to mention the many important. In addition there are more than 1,600 manuscripts called Lectionaries.

This is where six books were left out of the bible dealing with women's role in the church for the purpose of keeping women as sex objects and slave. Mary Magdalene was the holy grail. She carried the blood of Christ in his two sons who fathered the royal families of England and Europe.

There are 27 books and 200 chapters in the New Testament written early in the second century.

For centuries the major translation of the Bible was the Vulgate, a translation into Latin by the Bishop of Rome, Jerome. He worked 25 years doing this in Palestine and Bethlehem. It is still the official text of the Roman Catholic church. It was completed 440AD.

By 700 AD a monk known as the Venerable Bede began a translation of the Latin Vulgate into Anglo-Saxon. It was completed 735 AD. Earlier in the seventh century, Caedmon, a Saxon put some of the Bible's stories into song.

In 1522 Martin Luther translated the Old Testament and the whole Bible in 1534. He used every day expressions to make the meaning clear and come alive. He translated it from Hebrew and Greek, not the Latin Vulgate.

Martin Luther and other scholars were aided by Erasmus who produced a Greek New testament in 1516 with a parallel Latin translation.

In England, William Tyndale produced the Wycliffe Bible which was opposed by church authorities. Copies of it were hand written in secret and distributed by followers of Wycliffe who traveled all over England reading parts of it to people. Tyndale's translation was eventually printed and in 1536 he was martyred. His last words were, "Lord, open the King of England's eyes".

John Gutenberg invented printing the same year of 1536 printed Tyndale's New Testament in English, in Germany because it was forbidden in England.

Miles Coverdale finished the old Testament Tyndale had begun and revised Tyndale's New Testament. Henry VIII authorized it 1537. This Bible was chained inside the churches and was quite large. The Mathew Bible set up the basic text for the 'Great Bible', The Bishop's bible of 1568, the King James Bible, and most subsequent versions were of this text.

When Mary Tudor came to the throne, public use of the bible was again prohibited. Scholars fled to Geneva and undertook a new version of the bible. After a study of many Greek and Hebrew manuscripts, the Geneva Bible was published as a small book for use by the people; eventually brought to America on the mayflower in 1620 by Governor Bradford of the Plymouth Colony.

At a conference called by King James in 1604 Dr. John Reynolds proposed a translation that would have the approval of the whole Church. A commission of 54 scholars was appointed to undertake such a revision. Within the next 50 years it became the most popular version in the home.

When the United States became a nation in 1776 Bibles still had to be imported from England and Holland and were difficult to obtain. Bobert Atkins, a Philadelphia printer produced the first English Bible in America. It was commended to the public by a special Congressional Resolution. The Catholic bible includes some additional books that are not viewed as canonical by Jews and Protestants. These form the Apocrypha.

Other ancient writing not used in translating the Bible is the newly discovered Quam Ron Dead Sea Scrolls, the Emerald Tablet and thousands of ancient Sumerian writings. There are two others that I recently discovered and am trying to get a copy of.

In India there are thousands of books predating the Old Testament by many thousands of years. Some of them are in a language the nobody can read. Some are located in the London Museum.

One book that I would like to have is the Upanishad. It is Vedic literature dealing with the origins of the Universe and the coming of man.

The Cather's possessed a secret book of Love written by Jesus who gave it to john the Devine. The book of Love (Mari) or Amore (opposite of Rome) was lost when the Catholic Church intent on stamping out all other religions subjected the Cather's and Templar Knights (in 1308) to torture.

PS. I took courses in Man's Religions at University of Alaska. Islam and Christianity are two of the bloodiest religions on the planet except for communism that is responsible for the deaths of over 200-millino people.

Buddhism, Hinduism and Shintoism are much more benign religions respecting not only Gods laws but the laws of nature as well. The universe is constructed in such a way using ORMUS that responds to man's thoughts and prayer.

TUXEDNI BAY IS MY HOME

Tuxedni Bay is a pristine, unspoiled, wilderness and one of the few places left in the world where there is still abundant fish and game. It also has gold, platinum, clams, moose, bear, wolves, wolverine, otter and many other animals such as martin, weasel and beaver. It is sixteen miles long and five miles wide with tremendous views of ten-thousand-foot, Illiamna to the south and ten-thousand- foot, Redoubt to the north. Chisik Island in the middle of the entrance of the bay is 2,800 feet high. It shelters the bay from ocean swells. Thousands of saltwater, ducks and sea birds nest on the island because runs of sardines and needlefish come each spring. There are no roads to Tuxedni Bay and no phone lines. I have two airports plus the beach that I use to take off and land.

These products help people. This book is about helping people to get well and live long, healthy, pain-free, lives. It about discovering the Holy Grail, the Gold of the Gods and using it for good. This is an awesome responsibility. The underlying principal of the Grail monarchs was service in accordance with the Messianic code established by Jesus when he washed the feet of his Apostles' at the Last Supper. The right way to make money is by helping others.

When the Tulle Society met in Munich, Germany? They discussed: "...'a vial of black rock', ancient flying machines, and the invisible power source, the black sun." I eat a pinch of the black iridium, sands once in a while because I am convinced it is a beneficial source of many trace elements and monatomic minerals.

We have brown-bear-viewing, halibut fishing, salmon fishing, and bird watching, tours from an open skiff. The price of the tours is $50 for six hours. We can help people save thousands of dollars and decades of trial and error learning if you are planning to live in the wilderness.

There are free book downloads and games my web site: Philosophers Stone contains six recipes developed by Pfizer Pharmaceutical. My web site: www.GuardDogBooks.com. Click on the picture of the book. It will take you to the publisher.

There several free books on my web site you can download. One is a recipe book written in 1878 by 500 women. www.GuardDogBooks.com

Henry Kroll

26571 Heavy down drive

Soldotna, Alaska 99669
Or log onto: www.GuardDogbooks.com

We Made the Greatest Scientific Discovery of all time...

My friend Brad Guth and I are writing a book titled Cosmological Ice Ages. In it there is a chapter about the alleged: Global Warming. We charted the paths of the thirty stars in our local group looking for a grow light out there responsible for mopping up all the co2 every 105,000 years and we found one. It also does a whole lot more with regard to altering our life here on earth such as creating free oxygen, limestone, crude oil, natural gas and coal on land. These are the carbon-based resources we depend on for housing and energy. To find out more go: www.GuardDogBooks.com

ABOUT THE AUTHOR

Henry Kroll is the son of the Alaska's legendary, Mad Trapper. He was born on the family-owned floating cannery February 21, 1944 in the coastal fishing village of Seldovia, Alaska.

Henry learned to handle a rowboat at age five and by age sixteen was captain of the family, 50-foot power scow, Shrimp. At age twenty-two he and his wife purchased a boat named OUR LADY OF THE RIVERS in Chicago and traveled the length of the Mississippi to New Orleans and via the Inter-coastal Waterway to Brownsville Texas.

After attending Sheldon Jackson Jr. College, University of Alaska, and University of Corpus Christi, Texas he decided to fish king crab. After his father passed away Henry converted his father's power scow, Shrimp to fish the heavy 750 pound king crab pots. Together with his brother Herb Kroll they fished king crab in Katchemak Bay delivering crab to the town of Seldovia. Three years later the Mary M came up for sale.

Henry purchased the seventy-six-foot Mary M in Cordova, Alaska and converted it to fish king crab. Twenty-three years of handling 750 pound king crab pots in the icy waters of Cook Inlet, Kodiak and the Bearing Sea gave him a unique perspective that no other writer can match. During the 1980's he owned two large vessels, the 128 foot Orion and the 150-foot Victoria. Henry did all his own electrical, engineering and machine work.

As a commercial fisherman he caught over two-million pounds of salmon, three million pounds of king crab, three million pounds of tanner crab and several hundred thousand-pounds of halibut and shrimp. During his life on the sea he accumulated over twelve-thousand-days of Coast Guard documented sea time. He also ferried yachts for their rich owners from California to Seattle and Alaska earning his masters' license all oceans.

He credits his skill as a navigator for keeping him alive both in the air and on the sea.

In college Henry has worked as a radio announcer. He also mined gold, hunted treasure and built several small boats. Henry worked as a marine electrician, iron worker and welded everything. He plays piano for schools and churches. Currently he is doing motivational speaking and lecturing on the white-powder-of-gold, the MFKZT powder carried in the Ark of the Covenant.

Henry started writing in 1988 and now has twelve books to his credit. His unique ability to cause the reader to sense being a part of the situation, and of the solution, is mesmeric and he has a reputation for thorough and accurate coverage of subject matter.

Henry has five adult children and ten grandchildren. He and his wife, Mary are building a lodge in beautiful Tuxedni Bay located on the west side of Cook Inlet 180 miles southwest of Anchorage. Henry's scientific research on the inter-dimensional, monatomic iridium beach sands of Tuxedni Bay resulted in his writing, Philosopher's Stone.

Henry occasionally travels throughout the United States and does guest speaking about political issues, monatomic gold, ancient civilizations, and higher-dimensional spaceships. He will speak at radio stations, UFO conferences, schools and universities. If you have contacts with your local radio station, please contact us. Henry will play the piano for any audience.

His latest book, co-authored with Brad Guth, may be the greatest scientific discovery of the century titled: COSMOLOGICAL ICE AGES. They discovered that our sun orbits a two-solar-mass binary star with a white dwarf that puts out more than 100-times the light of our sun in the invisible UV spectrum responsible for creating the carbon energy on earth.

Our sun orbits just close enough to Sirius A and B to take Earth out of the ice ages. The invisible high intensity ultraviolet light from the white dwarf mops up CO_2 with massive plant growth both on land and in the world's oceans with diatoms creating limestone layers up to 12,500-feet thick and vast pools of crude oil. Each layer of coal on land represents one orbit around the Sirius binary star system. When you drive a car you are recycling stellar energy that did not come from the sun. Early earth's CO2 atmosphere was too thick to see the sun. You can't drive a car without a neutron star.

According to the American Indians we are now living in the fifth world. This means that there were four other advanced civilizations before us which were exterminated either due to natural events such as ice ages or downright stupidity. It is tantamount that we develop space travel to leave a historical record on other planets. If we continue to plod along at our present pace of scientific advancement there may be no one left to tell the story. The Permian Extinction wiped out practically all life on Earth right down to bacteria. Evolution did not produce the thirty-million or so plants and animals on the earth at the present time. Obviously someone is going around seeding planets between extinctions.

Brad Guth, one of the smartest men on the planet I am sorry to say passed away five years ago. He was one of the smartest men on the planet. After doing much research for our book Brad and I am now convinced that planet Nibiru could actually be located in the Sirius system. When we were in a close orbit to Sirius A and B Enki could have come to Earth in a spaceship filled with water as described in Zachariah Sitchin's translations of Sumerian Scrolls, in a voyage lasting eight years. His brother Enlil came later to get gold to turn it into the white powder. They needed the ORME antigravity gold to spray in the atmosphere of their planet to protect it from the harmful rays of Sirius B emitting more than 100-times the light of our sun in the ultraviolet 350 to 400 nanometer wave length. The Orme gold also made the people living on Nibiru live hundreds of years and make all living things including plants thrive.

We discovered that the carbon energy on Earth wasn't made by the sun. Early Earth's carbon CO2 atmosphere had to be over 1000 pounds per square inch. Such an atmosphere would be 2,800 miles deep.

Right now it is only 50 miles deep and 14.5 Psi at sea level. Earth's expansion with plate tectonics and 40-thousand tons a year of incoming meteorites caused earth to double I size thereby reducing the atmospheric press by half. The rest was laid down as coal, oil and limestone by photosynthesis with light that could not have come from the sun because the atmosphere was still over a thousand miles deep during the Carboniferous Era.

The sun did not make the coal, oil or limestone layers up to 12,500-feet thick because the atmosphere was about 2000 miles deep. You probably wouldn't see the sun with such a thick atmosphere. The reason you have oil to drive your car was our solar system was captured by the Sirius System.

I discovered that our sun was born in Orion and our fortunate capture by the Sirius binary star system furnished the light to make most all of the energy on this planet, not the sun!

Little Sirius B, the size of earth, puts out more than 100 time the light of our sun and when it orbits close to Sirius A could put out 1000 times the light of our sun in the invisible UV spectrum slightly above the range of human sight in the 360 to 300 nanometer range. It was the light from these objects not the sun that made most all the coal, oil and limestone on earth. It was their light that allows you to drive a car not the sun!!! Humanity has been worshiping the wrong sun god for 4000 years and it is no wonder since earth only recently came out of the Ice Age.

NATURAL SUPPLIMENTS

Trees have evolved over hundreds of millions of years to where their sap or resin had to ward off insects, mold, bacteria, and other parasites. When you take the sap of a birch or maple tree and boil it down ten to a hundred to one you will be concentrating the tannins, insecticide and psychoactive chemicals in the sap. Some, not all, of these chemicals may be useful in restricting or even curing some, not all diseases. Below is a redacted study of maple syrup. I urge you to look up and read the complete study.

There is a study out describing how maple syrup slows and sometimes destroys cancer cells especially colorectal cancer.

Chaga mushroom is a black or brown, woody-parasite that grows on the bark of birch trees feeding on the sap. When copped off with a hatchet and either ground up or chopped into chunks it is put in boiling water to make a medicinal tea.

Cancer cells cannot grow in an alkaline body. Some doctors and ordinary people put a level teaspoon of baking soda in a glass of water and drink it every day to alkalize the body in the hope it will ward off cancer. It also has been used of upset stomach for many years.

The theory I discovered in how to slow or possibly cure colorectal cancer using chaga mushroom and maple syrup is as follows: Put a level or quarter teaspoon of baking soda in a cup of boiling water, add a heaping teaspoon of ground chaga and sweeten with dark maple syrup. It actually tastes good!

The theory about how it works is this: cancer cells love sugar. They gobble up the sugar and when they ingest the baking soda and chaga they either explode or die.

You can now buy ORMUS on Amazon.

The philosopher's stone or manna—also known as the Hebrew MFKZT powder or white-powder-of-gold and m-state was carried inside the Ark of the Covenant. If you were to peek inside you would see a glowing white powder with little lightning bolts trickling over its surface. This antigravity, room temperature, superconductor is used for cell division by all living things and was used by God's angels to create man.

David Hudson spoke in a lecture: "The manna or the elixir of life when mixed with water forms a gelatinous mixture. When ingested it will have the following affects; every cell in your body will be taken back to the state it is supposed to be, when you were a teenager or a child. It perfects the DNA, and closes the light within the body until you literally reach a point where the light body exceeds the physical body." "The gifts that go with this are perfect telepathy, you can know good and evil when it is in the room with you, you can project your thoughts into someone else's mind, you can levitate, you can walk on water, because it is flowing so much light in you, you literally don't attract to gravity."

Teams of scientists in Japan, Germany, China, and the US are currently trying to get a jump-start on this technology. It was recently discovered in the air we breathe and the water we drink. The air filters, which you plug into a wall outlet in your home to collect dirt, dust and pollen also, collect ORME atoms. It looks like a dirty-white gel. When you touch it with your finger it disappears because the magnetic field of your finger puts it into a higher energy state. This may be the substance the answers the Parana energy mystery. Athletes can lift more and run faster after taking a deep breath. It can't be due to the oxygen intake because oxygen takes several minutes to get into the blood stream. The human olfactory nerves in the nasal passage seem to have the ability to absorb ORME atoms.

This material super-conducts magnetic fields during cell division and corrects the DNA in all living things. It can extend your life hundreds of years and increase you IQ hundreds of points. It is the only known room-temperature superconductor. It repels the force of a magnet! It improves electricity production in fuel cells and solar cells. The only limits to its uses are your imagination and it can expand your mind so that too is unlimited. The physical shape of these atoms themselves give it a Meissonier frequency similar to brain waves. In fact, this material may furnish the power to operate your brain. The signal being transmitted instantly connects all life forms throughout the Universe. When you eat more of the stuff this connection becomes much stronger!

When used to bring forth the Shekinah glory (electric discharge of holy fire) over the top of the Ark a signal can be broadcast across the galaxy in one Planck second. You don't have to wait millions of light years for an answer when communicating with other intelligent races in our galaxy. If you tripled your intelligence you could instantaneously produce holographic thought-form, images above an identical Ark on the other side of the Galaxy on whichever planet this device happens to be operational. Knowledge of the ORME (Orbitally Rearranged Monatomic Elements) particle which Moses's metal smith, Bazaleal create by burning gold at 5600 degrees is a little difficult for people to understand. The concept of a monatomic particle that cannot bond into a solid substance because all it's electrons are orbiting in pairs, a particle with a magnetic frequency of brain waves enabling the observer to manipulate the outcome of an experiment or event. This advanced; alien-mind-control science is beyond our comprehension.

It exists, as a glowing white power in its pure state with thousands of amps of electricity flowing through it but no voltage unless excited by brain waves or some other outside magnetic frequency. This is the primal atom used to regulate the outcome of a singularity during the creation of the Universe. It weighs five-eight's the weight of the actual metal itself at room temperature but when heated weighs less than nothing and disappears from third dimension. When it cools it reappears again from a higher dimension. The ancient Egyptian translation of manna is "What is it?" It is the food of the Gods and was used to take mankind into higher dimensions where time and distance are inconsequential. This has been done since the beginning of time.

You eat it in your food every day yet most people are completely unaware that it exists. People pray over their food to energize it so that it will nourish their body more completely and this material actually responds to those prayers by altering the frequency thereby making the food more compatible and digestible. It is the light of life and absolutely necessary for cell division in all living things. Plants take it out of the soil by dissolving it with alkaline extruded from their roots. You break the plants down using hydrochloric acid in your stomach. The reason why some plants are poison is because they contain too much alkaline. Most plant poisons are alkaline poisons.

The use of this material makes the cloning of all life forms possible. When used as such it gives mankind the ability of a God to speed up evolution. Such a power is I feel in accord with God's will when used for a worthwhile cause such as getting us off the planet to assist and create life in other biosphere's. The use of this material will also reduce the stress on our own biosphere. We must take this next evolutionary step or perish.

For more information on this subject read my book on astronomy titled: COSMOLOGICAL ICE AGES available hankkroll.com. Also check out Home Of The Angels and other books.